VACCINES

From Concept to Clinic

A guide to the
development
and clinical testing
of vaccines for
human use

Edited by

Lawrence C. Paoletti
Pamela M. McInnes

CRC Press
Taylor & Francis Group
Boca Raton London New York

CRC Press is an imprint of the
Taylor & Francis Group, an **informa** business

CRC Press
Taylor & Francis Group
6000 Broken Sound Parkway NW, Suite 300
Boca Raton, FL 33487-2742

First issued in paperback 2019

© 2008 by Taylor & Francis Group, LLC
CRC Press is an imprint of Taylor & Francis Group, an Informa business

No claim to original U.S. Government works

ISBN-13: 978-0-8493-1168-0 (hbk)
ISBN-13: 978-0-367-40037-8 (pbk)

Library of Congress Cataloging-in-Publication Data

Vaccines, from concept to clinic : a guide to the development and clinical testing of vaccines for human
use / edited by Lawrence C. Paoletti and Pamela M. McInnes.
    p. ; cm.
Includes bibliographical references and index.
ISBN-13: 978-0-8493-1168-0 (hardcover : alk. paper)
ISBN-10: 0-8493-1168-3 (hardcover : alk. paper)
    1. Vaccines. 2. Vaccines--Testing. I. Paoletti, Lawrence C. II. McInnes, Pamela M.
[DNLM: 1. Vaccines, Synthetic. 2. Research Design. 3. Clinical Trials. QW 805V11644 1998]
QR 189.V275 1998

65'.372--dc21
DNLM/DLC

98-21739 CIP

Visit the Taylor & Francis Web site at
http://www.taylorandfrancis.com

and the CRC Press Web site at
http://www.crcpress.com

Foreword

The field of vaccinology has expanded exponentially during the past 20 years. It is no longer merely a small branch of microbiology, immunology, or clinical infectious diseases but a complex and substantive discipline in its own right.

A number of high-profile events—the HIV epidemic, the increase in antimicrobial resistance of bacterial pathogens, the emergence and reemergence of infections from Ebola virus to diphtheria—all have underscored, for scientists and the public, the critical role that immunization must play if infectious diseases are to be successfully controlled over the long term.

This is truly an exciting time for this fledgling field as vaccinologists extend their scope from the prevention of the classic infections of childhood such as diphtheria, whooping cough, and measles, to the prevention of infections in all age groups and all over the world.

For example, vaccines are being developed against sexually transmitted diseases targeted for adolescents and adults, against respiratory viruses targeted to high-risk adults and the elderly, and against group B *Streptococcus* targeted to mothers for the protection of their babies. In addition to these vaccines for universal use, numerous agents are being developed for more specialized uses in particular geographic areas (e.g., dengue, malaria), in travelers or in high-risk groups such as surgical patients and the immunocompromised. With the advent of more powerful adjuvants and lymphokines which may allow us to both enhance and modulate the immune response, it may soon become possible to develop therapeutic vaccines to treat chronic or recurring infections such as hepatitis B, hepatitis C, and genital herpes. These technologies may ultimately allow vaccinologists to design vaccines for the immunotherapy of cancer, or for the control of autoimmune diseases and allergies.

There are numerous manifestations of the expanding interest in vaccine research. National and international meetings occur not on an annual but on a monthly or even weekly basis. Many medical schools and public health schools offer courses in vaccine development and evaluation. A number of academic institutions have founded vaccine research and/or clinical evaluation centers that are supported by national institutes of health and by industry. Several national and international institutes devoted to vaccine research and development have recently been created. Vaccine literature is burgeoning with numerous papers appearing in major scientific journals; two journals devoted exclusively to vaccines have appeared; and there are countless books and monographs on specific vaccines, vaccine technologies, adjuvants, vaccine delivery devices, and the like.

Vaccines: From Concept to Clinic fills an important gap in this extensive vaccine literature. Designed with academic vaccine researchers in mind, it provides a road map of how a vaccine is taken from an idea in a researcher's imagination to the lab bench through preclinical evaluation, into the clinic for safety immunogenicity and efficacy, and ultimately to commercialization as a licensed product.

Those who have traveled down this complex path during the past 20 years, and the editors Lawrence Paoletti and Pamela McInnes are among them, realize how much they would have benefited if this book had been on their shelf when they began.

The first portion of this book is devoted to the preclinical stages of vaccine development. The first chapter by Rino Rappuoli and Giuseppe Del Giudice provides a lucid and thoughtful description of how vaccine targets are identified and prioritized from both a public health and a commercial vantage and how approaches are selected rationally, from the many available, to develop a vaccine against a particular infectious disease.

In Chapter 2, Andrew Onderdonk and Ronald Kennedy review the use of animal models in the preclinical evaluation of vaccines, with special emphasis on primates. At the least, every candidate vaccine must be evaluated for safety, toxicity, and immunogenicity in appropriate animals before human studies are initiated. At best, animal models of infection can provide valuable insights into a vaccine's protective activity and its mechanism and thereby enhance the likelihood of its success in preventing human disease.

The third chapter, by Dace Madore, Nancy Strong, and Sally Quataert, thoroughly reviews the validation, standardization, and calibration of serologic assays, a critical issue in the preclinical and clinical evaluation of vaccine immunogenicity and ultimately in the development of serologic correlates of protection.

A practical guide for the academic researcher who wishes to prepare a vaccine for initial Phase I human evaluation is provided in the fourth chapter, by Lawrence Paoletti. Although it is personally rewarding for the academic investigator to take a vaccine idea into human trials, the regulatory requirements have become increasingly stringent, and thus the personal commitment and resource requirements have become truly daunting. Consequently, most academic scientists now link up with an established public or private sector vaccine development laboratory to accomplish this task.

The middle portion of the book deals with rules of the road in the process of clinical vaccine investigation from the vantages of both the driver (Chapter 5) and the traffic cop (Chapter 6). Martha Mattheis and Pamela McInnes provide a practical investigator's guide to writing an Investigational New Drug application, based on their own extensive experience at NIAID in writing such documents. In Chapter 6, Donna Chandler, Loris McVitty, and Jeanne Novack give the regulator's perspective of the IND process, including a valuable list of common pitfalls.

Jane Biddle describes the ins and outs of patenting technology, interpreting material transfer agreements, and negotiating licensing agreements in Chapter 7. Like physicians practicing under managed care systems, vaccine researchers also have had to become businesspersons in order to succeed in getting their vaccines developed. Familiarity with the technical details and rules of the road for this transition from the academic laboratory to the commercial laboratory is absolutely essential for academic vaccine researchers.

The final two chapters of the book are both entertaining and inspiring sagas of the development of two vaccines, a malaria vaccine in David Kaslow's Chapter 8 that is still a "work in progress," and the Oka strain varicella vaccine in Michiaki Takahashi's Chapter 9 that is now marketed globally. The common thread in both stories is that they amply illustrate the three most important character traits of the successful vaccine developer—persistence, persistence, and persistence.

George R. Siber, M.D.
Pearl River, New York

Preface

The scientific literature today is replete with books on vaccines, written and edited by experts active in the field. Recently published books review the state of research on vaccines against the most common, as well as the most rarely encountered, diseases of humans and animals. Instructive books and manuals describe new methods of vaccine preparation, delivery, handling, and storage. Monographs from domestic and international vaccine meetings, information originating from several academic sources, and audiovisual lectures from meetings are readily accessible via the Internet. An international peer-reviewed journal dedicated to this field completes the information base. So why initiate another book on vaccines? What else can be added to the vaccine field that has not yet been covered?

This book arose as we began to generate a clinical lot of polysaccharide-protein conjugate vaccine against group B *Streptococcus*. For the group B streptococcal vaccine project, research conducted in the 1970s and 1980s told us what we needed to do technically in the laboratory to create the vaccines; however, information on the steps necessary to manufacture, bottle, and test clinical lots of this vaccine was not readily available. This information was gleaned primarily from conversations and correspondence with colleagues in industry and government. We realized that compiling the necessary information in one book might be useful to researchers who were on the verge of producing a lot of vaccine for Phase I clinical trials and who were asking these questions: Can I produce a clinical lot of vaccine in my laboratory? How should the vaccine be vialed? What are the expectations of the FDA? What is an IND application and how is it filed? What are CFRs? Which CFRs apply to the production of a vaccine? What is potency and how is it measured?

We learned that the answers to these questions resided in publications from different regulatory agencies and in the minds and notebooks of those who had gone through the process of manufacturing a vaccine for clinical use, filing an IND application, and designing and performing clinical trials. Without access to such resources, the young vaccine developer could be bewildered and frustrated by the complexity of the endeavor. Through this book we seek to provide guidance in this process. The core issues around which the rest of the book was developed are: considerations for the preparation of a clinical lot of vaccine, an understanding of each phase of clinical testing, knowledge of the organization and preparation of a manufacturer's protocol and an IND application, and awareness of the concerns of those who review INDs. Areas considered critical to vaccine development—identification of protective epitopes, development of animal models or test systems, appropriate design and performance of serological assays—are at the front of the book. Because technology transfer has become a central concern of scientists whose discoveries may have not only a public health impact but also an economic impact, we include a chapter on this critical but often confusing topic as a reference for both the novice and the experienced scientist. Students of vaccine research may gain invaluable insights into vaccine research and development by reading the final segment of the book, which describes the personal trials and tribulations encountered in the development of two vaccines: one that has gained widespread acceptance for human use (varicella) and one that is still in its developmental stages (malaria).

This book would not have been created without the commitment of the authors and the encouragement of our colleague, Dr. Arthur Tzianabos. We thank Jaylyn Olivo and Julie McCoy of the Brigham and Women's Hospital Editorial Service for reviewing each chapter (their talents in this field are without equal). We also acknowledge with gratitude those at our respective institutions who have created and nurtured an environment that allows the pursuit of this kind of project.

Lawrence C. Paoletti and Pamela M. McInnes

About the Editors

Lawrence C. Paoletti

Dr. Paoletti graduated in 1988 from the University of New Hampshire at Durham with a doctoral degree in Microbiology. He joined the Channing Laboratory, Brigham and Women's Hospital, Boston, to pursue postdoctoral studies on the development of oligosaccharide-protein conjugate vaccines against group B streptococcal disease. In 1991, Dr. Paoletti became part of a team, which included members of the NIH's NIAID, developing the first group B streptococcal capsular polysaccharide-protein conjugate vaccines for use in clinical trials. Several other group B streptococcal conjugate vaccines are now in various stages of clinical evaluation. His experiences in the academic pursuit of producing vaccines for clinical trials provided the inspiration for this book.

Pamela M. McInnes

Dr. McInnes is currently Chief of the Respiratory Diseases Branch, Division of Microbiology and Infectious Diseases, National Institute of Allergy and Infectious Diseases, National Institutes of Health. Before taking this position, she served as the Neonatal Pathogens and Maternal Immunization Program Officer in the Respiratory Diseases Branch, having joined the Respiratory Diseases Branch in 1990, after a year as an NIH Grants Associate. Previously, Dr. McInnes was Associate Professor of Orthodontics at LSU School of Dentistry, in New Orleans, Louisiana. She received her dental degree and advanced degree in biomaterials from the University of the Witwatersrand in Johannesburg, South Africa.

Dr. McInnes is actively involved with the NIH extramural research community in the mission of reducing the morbidity and mortality attributable to respiratory diseases. She has a particular interest in vaccine development and clinical evaluation.

Contributors

Jane A. Biddle, Ph.D.
Jane A Biddle Consultants
4209 West Bertona St.
Seattle, WA 98199

Donna K.F. Chandler, Ph.D.
Division of Vaccines and
  Related Products Applications
Office of Vaccines Research
  and Review
Center for Biologics Evaluation
  and Research
Food and Drug Administration
1401 Rockville Pike
Rockville, MD 20852-1448

Giuseppe Del Giudice, Ph.D.
IRIS, The Chiron-Vaccines
  Immunobiological Research
  Institute in Siena
Via Fiorentina 1
53100 Siena
Italy

David C. Kaslow, M.D.
Head, Malaria Vaccines Section
  and Recombinant Protein
  Expression Unit
Laboratory of Parasitic Diseases
Building 4, Room B1-31
NIAID, NIH
Bethesda, MD 20892-0425

Ronald C. Kennedy, Ph.D.
Department of Microbiology and
  Immunology
The University of Oklahoma
  Health Sciences Center
P.O. Box 26901
Oklahoma City, OK 73109

Dace V. Madore, Ph.D.
Director
Immunobiological Services
Wyeth-Lederle Vaccines
  and Pediatrics
211 Bailey Road
West Henrietta, NY 14586-9728

Martha J. Mattheis, M.S.,
Chief
Clinical and Regulatory
  Affairs Branch
Division of Microbiology and
  Infectious Diseases
NIAID, NIH
Solar Building
Room 3A01
Bethesda, MD 20892-7630

Pamela M. McInnes, D.D.S.,
  M.Sc. (Dent.)
Chief
Respiratory Diseases Branch
Division of Microbiology and
  Infectious Diseases
NIAID, NIH
Solar Building, Room 3B04
Bethesda, MD 20892-7630

Loris D. McVittie, Ph.D.
Division of Vaccines and
  Related Products Applications
Office of Vaccines Research
  and Review
Center for Biologics Evaluation
  and Research
Food and Drug Administration
1401 Rockville Pike
Rockville, MD 20852-1448

Jeanne M. Novak, Ph.D.
Division of Vaccines and Related
  Products Applications
Office of Vaccines Research and
  Review
Center for Biologics Evaluation
  and Research
Food and Drug Administration
1401 Rockville Pike
Rockville, MD 20852-1448

Andrew B. Onderdonk, Ph.D.
Channing Laboratory
Brigham and Women's Hospital
Harvard Medical School
181 Longwood Avenue
Boston, MA 02115

Lawrence C. Paoletti, Ph.D.
Channing Laboratory
Brigham and Women's Hospital
Harvard Medical School
181 Longwood Avenue
Boston, MA 02115

Sally A. Quataert, Ph.D.
Wyeth-Lederle Vaccines and
  Pediatrics
211 Bailey Road
West Henrietta, NY 14586-9728

Rino Rappuoli, Ph.D.
IRIS, The Chiron-Vaccines
  Immunobiological Research
  Institute in Siena
Via Fiorentina 1
53100 Siena
Italy

George Siber, M.D.
Director of Research
Wyeth-Lederle Vaccines and
  Pediatrics
401 N. Middletown Road
Pearl River, NY 10965

Nancy M. Strong
Wyeth-Lederle Vaccines and
  Pediatrics
211 Bailey Road
West Henrietta, NY 14586-9728

Michiaki Takahashi, M.D.
Emeritus Professor of Osaka
  University
Director, The Research Foundation
  for Microbial Diseases of Osaka
  University
3-1 Yamada-Oka, Suita
Osaka, 565
Japan

For the children

Table of Contents

1 Identification of Vaccine Targets

Rino Rappuoli and Giuseppe Del Giudice

INTRODUCTION

The identification of vaccine targets is a complex endeavor that requires a multidisciplinary approach that involves epidemiological analyses to identify the diseases for which vaccines are needed, a market analysis to verify whether the developed vaccines will be commercially viable, and a feasibility study to ascertain whether the vaccine development is feasible with the technologies and the knowledge available. In this chapter, we will analyze each of these steps and provide examples of how they can be approached.

IDENTIFICATION OF THE DISEASES FOR WHICH VACCINES ARE NEEDED

The only rational approach to the assessment of vaccine need can be provided by epidemiological considerations. Data on the incidence and prevalence of known infectious diseases and on emerging infectious diseases are regularly published and referenced by several agencies, including the World Health Organization,[1] the World Bank,[2] the National Institute of Allergy and Infectious Diseases,[3] and the Centers for Disease Control and Prevention.[4] These data are usually not complete, for they reflect mostly the situation in countries with good epidemiological surveillance, while they underestimate the needs of countries with poor surveillance. For instance, most of the data available reflect the incidence and/or prevalence of communicable diseases in the United States, and, to a lesser extent, those in the European countries. Nevertheless, even if limited, these data are an important source of information. A summary of the most prevalent infectious diseases and their incidence is reported in Table 1.1.

However, the epidemiological considerations alone are not enough to decide the priorities for vaccine development. Technical feasibility and economic return on the investment are also important. Of the diseases reported in Table 1.1,

0-8493-1168-3/99/$0.00+$.50
© 1999 by CRC Press LLC

Table 1.1. Worldwide Prevalence of Major Infectious Diseases[a]

Disease	Cases (Prevalence) (million)	Deaths/year (million)
Diarrhea	4000	3.1
Malaria	500	2.1
Hepatitis B	350	1.1
Ascariasis	250	0.06
Schistosomiases	200	0.02
Filariasis	120	—
Hepatitis C	100	—
Gonorrhea	60	—
Measles	42	1.0
Pertussis	40	0.35
HIV	40	1.0
Tuberculosis	22	3.0
Leishmaniases	12	0.08
Cholera	0.4	0.011

[a] Adapted from Reference 2.

only a minority have a high priority for vaccine development, primarily because of two factors: a) many of the most prevalent diseases are caused by parasites for which vaccine development has been largely unsuccessful with the technologies available up to now, and b) most of these diseases affect developing countries, with a consequent low economic interest of northern countries in the development of these vaccines. The agents against which active vaccine development is currently underway are reported in Table 1.2.

TYPES OF VACCINES

Existing vaccines can be divided into three broad categories, depending on whether they contain live attenuated microorganisms, inactivated whole microorganisms, or purified components of microorganisms (subunit vaccines). A new, popular category of vaccines is now represented by the nucleic acids vaccines.

Live Attenuated Vaccines

Examples of widely used live attenuated vaccines include the BCG vaccine against tuberculosis;[5-6] the Sabin type of polio vaccine;[7] the vaccines against measles, mumps, rubella,[8] and varicella;[9-10] and the Ty21A vaccine against typhoid fever.[11] Most of these vaccines were developed before the era of molecular biology and biotechnology by *in vitro* passage of human pathogens; therefore, the molecular mechanisms of their inactivation are unknown. Today such noncharacterized vaccines would be difficult to introduce. However, well-characterized, live attenuated bacterial and viral strains can be built by rational modification of the genome of the pathogen. Thus far, most of the work has been

Table 1.2. Stage of Development of Vaccines Against Several Infectious Agents and Effort Spent in the Development[a]

Target Agent	Stage of Development	Effort
Bordetella pertussis	Phase III, Licensed	very high
Borrelia burgdoferi	Basic R&D	medium
Brugia malayi	Basic R&D	low
Chlamydia	Basic R&D	medium
Coccidioides immitis	Basic R&D, Phase III	low
Cryptococcus neoformans	Basic R&D	low
Cytomegalovirus	Phase I	high
Dengue virus	Basic R&D, Phase I	high
Entamoeba histolytica	Basic R&D, Phase II	low
Enterotoxinogenic *E. coli*	Phase II	high
Epstein-Barr virus	Phase I	medium
Group A *Streptococcus*	Basic R&D, Phase I	medium
Group B *Streptococcus*	Phase II	medium
Hepatitis C virus	Basic R&D, Phase I	high
Hepatitis D virus	Basic R&D	low
Hepatitis E virus	Basic R&D	low
Helicobacter pylori	Basic R&D, Phase I	medium
Herpes simplex virus	Phase III	high
Histoplasma capsulatum	Basic R&D	low
HIV	Basic R&D, Phase II	very high
Human papilloma virus	Phase I	high
Influenza virus	Phase I, Phase II	high
Legionella pneumophila	Basic R&D	low
Leishmania	Basic R&D, Phase III	low
Measles virus	Basic R&D, Phase III, Licensed	medium
Mycobacterium leprae	Basic R&D, Phase III	low
Mycobacterium tuberculosis	Basic R&D	high
Mycoplasma pneumoniae	Basic R&D	low
Neisseria gonorrheae	Basic R&D	medium
N. meningitidis A	Phase II	high
N. meningitidis B	Basic R&D, Phase I	high
N. meningitidis C	Phase II	high
Parainfluenza virus	Phase I, Phase II	medium
Plasmodium	Basic R&D, Phase I-III	high
Pseudomonas	Basic R&D, Phase I	medium
Rabies virus	Phase III, Licensed	medium
Respiratory syncytial virus	Phase I, Phase II	medium
Rotavirus	Phase I-III	high
Salmonella	Phase I-III, Licensed	high
Schistosoma	Basic R&D	medium
Shigella	Phase I-II	medium
Streptococcus pneumoniae	Phase III	high
Toxoplasma gondii	Basic R&D	low
Treponema pallidum	Basic R&D	low
Vibrio cholerae	Basic R&D, Phase I, Phase III	high

[a] Adapted from Reference 3.

dedicated to obtaining attenuated strains of, for example, *Salmonella* by deleting or inactivating the genes coding for the synthesis of aromatic amino acids or components of regulatory pathways,[12] *Vibrio cholerae* by deleting the gene coding for the A subunit of cholera toxin,[13–14] and vaccinia virus by deleting genes involved in nucleotide metabolism.[15–16] However, in theory, any microorganism can be attenuated by deleting or modifying genes that are essential for the *in vivo* growth of the pathogen.

Vector-Based Vaccines

Today, the most attractive reason to develop live attenuated microorganisms resides in their ability, acquired through genetic engineering, to produce *in vivo* cloned antigens derived from other microorganisms. A great deal of literature is available describing *Salmonella* spp., poxviruses, and many other attenuated microorganisms that express recombinant antigens.[17–19] This approach can be useful when we want to induce mucosal immunity and cytotoxic responses to target antigens.

Inactivated Vaccines

Heat or chemical inactivation of bacteria and viruses has been the first, easy approach to vaccines. Its advantage is that all antigens present in the pathogen are included, so that it is not necessary to know which are the protective antigens. The disadvantage is that some of the vaccine components may be toxic and responsible for side effects. Today, this method of vaccine development is no longer common, although several vaccines of this type are still widely used: the whole-cell vaccine against pertussis,[20] the Salk polio vaccine,[21] the influenza vaccine,[22] and the vaccines against rabies[23] and tick-borne encephalitis.[24]

Subunit Vaccines

Subunit vaccines consist of one or more antigens, purified from the microorganism or produced by recombinant DNA technology, that are able to protect against the disease. Development of subunit vaccines requires knowledge of the protective antigen(s) and the ability to produce and purify them on a large scale. It is also desirable to know the type of immune response that will induce protection, in order to be able to construct and deliver the antigen in the appropriate way.

The first subunit vaccines to be developed have been diphtheria and tetanus toxoids.[25] In this case, the observation that both diseases were caused by a toxin produced by the bacterium suggested that serum antibodies able to neutralize the toxin were sufficient to protect from disease. Therefore, the semipurified toxins were inactivated by chemical (formaldehyde) treatment and used as vaccines.

A second example of subunit vaccines are polysaccharides and conjugated vaccines against encapsulated bacteria.[26-27] In this case, the observation that serum bactericidal antibodies against the capsular polysaccharide were enough to protect from invasive bacterial infection suggested the development of purified capsular polysaccharides as vaccines. Polysaccharide vaccines have been developed against *Neisseria meningitidis* serogroups A, C, Y, W135, against 23 types of *Streptococcus pneumoniae*, against *Haemophilus influenzae* type B, and against *Salmonella typhi*. However, polysaccharide vaccines elicit T cell-independent immune responses, inducing primarily IgM in adults, and no immunity at all in infants. To overcome these problems, capsular polysaccharides were covalently linked to carrier proteins, thus obtaining semisynthetic conjugated vaccines that induce T cell-dependent responses in both adults and infants.[28] The conjugate vaccine against *H. influenzae* type B has dramatically reduced the disease in those countries where it has been introduced. Many conjugate vaccines are now under development, such as those against *N. meningitidis* serogroups A and C,[29-30] *S. typhi*,[31] and *Streptococcus agalactiae* (group B *Streptococcus*).[32]

A third prototype subunit vaccine is the recombinant vaccine against hepatitis B.[33] In this case, it had been observed that serum antibodies elicited by the envelope protein of the hepatitis B virus were able to neutralize the virus and to protect from infection. However, the vaccine could not be produced initially in large quantities because the virus did not grow *in vitro* and could only be purified from the plasma of infected patients. Therefore, recombinant DNA was used to engineer a yeast strain to produce the envelope protein. This turned out to be one of the rare cases in which the yeast produced and assembled the protein in the correct conformation, so that the purified recombinant protein could be used directly as vaccine.

Another type of subunit vaccine is the recently developed acellular pertussis vaccine.[34] In this case, a considerable amount of work was initially required to identify the bacterial antigens that were able to induce protective immune responses. Pertussis toxin (PT) was identified as a major protective antigen; other antigens, such as adenylate cyclase, filamentous hemagglutinin, pertactin, and fimbriae, were found to contribute to protective immunity. Therefore, these antigens were used in different combinations in candidate vaccines.[35-37] The experimental rationale for selecting protective antigens will be described later in this chapter. In order to be introduced in the vaccine, PT was detoxified in several ways. Most vaccine preparations contained PT detoxified by a classic chemical treatment used since the beginning of the century to detoxify diphtheria and tetanus toxins. However, in one of the vaccines, the PT was detoxified by one of the powerful tools of molecular biology—site-directed mutagenesis—to stably change the amino acids responsible for the toxicity of PT. This molecular approach made it possible to obtain a naturally nontoxic molecule, inactivated by rational design, that did not need a denaturing chemical treatment and that showed superior immunogenicity and protective efficacy in clinical trials.[38-40] The recombinant acellular pertussis vaccine, the first example of a vaccine obtained by rational drug design, shows that this approach can provide tremendous advantages, eliciting highly protective immune responses with small amounts of antigen.

Nucleic Acid Vaccines

The use of naked DNA as an approach to the induction of immune responses is a new but popular method in vaccinology that is potentially of great importance.[41] In this approach, the gene(s) coding for the antigens against which we want to raise an immune response are cloned in appropriate plasmid vectors under strong promoters and directly injected into the host. A small fraction of the injected plasmid DNA is then taken up and expressed by antigen-presenting cells; the result is the elicitation of an immune response against the newly expressed foreign antigen. Although young, this technology has been experimentally applied to many vaccines, often with very promising results. Influenza, tuberculosis, and malaria, and infection with HIV, hepatitis B virus, and human papilloma virus, are just a few of the infections that are now being pursued by this fast-growing technology. The naked DNA approach has two great advantages over conventional technologies. First, it is very simple. The same, easy manufacturing technology (plasmid DNA preparation) is used for all antigens and all vaccines, so that once the protective antigen(s) have been identified and cloned, the vaccine development path is identical for all vaccines. Second, it is the most efficient method known to date for inducing a cytotoxic immune response against an antigen. Some unsolved questions are, however, still present: a) The safety issues of injecting DNA remain unsolved. b) Antigens are folded and posttranslationally modified by eukaryotic cells. For instance, bacterial antigens are often glycosylated and may not have the appropriate conformation when they are given as naked DNA molecules. c) This method cannot be applied to the generation of polysaccharide vaccines. d) The delivery of the DNA is very inefficient, and large quantities of plasmid are still needed to elicit strong immune responses.

Clinical trials over the next few years will confirm whether the promising results obtained so far in mice and primates lead to the development of a new class of vaccines providing protection against many of the diseases for which the classic technologies have failed or been inadequate. For instance, the possibility of immunization with many genes from the same pathogen may allow us to approach complex parasitic diseases such as malaria.

IDENTIFICATION AND SELECTION OF PROTECTIVE ANTIGENS

The identification of protective antigens is a complex problem that involves different approaches for different viruses, bacteria, and parasites. Viruses usually have a small genome that encodes a few proteins, and thus, the selection of antigens is simpler than in other microorganisms. Envelope proteins and glycoproteins are the primary candidates for induction of virus-neutralizing antibodies, while core antigens are usually good candidates for cytotoxic T cell responses.

In microorganisms with complex genomes, such as bacteria and parasites, there are several hundred possible candidate antigens; therefore, the identification of protective antigens requires a rational approach that often combines genetic, biochemical, and immunological studies. A good starting point is the study of pathogenic mechanisms. It is often possible to isolate or construct forms of the microorganism that are nonpathogenic because they do not produce one or more molecules. In these cases, the molecules that are made only by the pathogenic microbes, for example, bacterial toxins and capsular polysaccharides, are the basis for many successful vaccines. However, in several instances the situation may be more complex, as is the case for *Bordetella pertussis* and *Helicobacter pylori*, as described below.

The experience with the *B. pertussis* whole-cell vaccine, which contained heat-killed bacteria, had clearly shown that vaccination with all pertussis antigens conferred protection from disease. To develop acellular vaccines composed of a few purified components, it was necessary to identify among the several hundred antigens contained in the whole-cell vaccine only those that were able to confer protection and to eliminate those that were useless or toxic.[37] The first approach to the identification of protective antigens was the observation that, while whole-cell vaccines made by *B. pertussis* grown at 37°C conferred good protection in the animal model, whole-cell vaccines containing bacteria grown at 25°C or in the presence of 50 mM $MgSO_4$ were unable to induce protective immunity. Biochemical and genetic studies showed that bacteria grown at 37°C produced several antigens that were not produced at 25°C. Therefore, the search for protective antigens was restricted to those that were produced at 37°C. Excision of some of the genes expressed only at 37°C showed that only a minority of them were essential for virulence.[42-43] Among these was the gene for pertussis toxin, a molecule that in a detoxified form became the main component of all acellular pertussis vaccines. The other vaccine components, such as filamentous hemagglutinin, fimbriae, and pertactin, were selected among those expressed only at 37°C as being easy to purify and participating in the induction of protection in the animal model.

In the case of *H. pylori*, two approaches have been followed. The first is the selection of the most abundant antigen (the urease), which is easy to purify and to make in a recombinant form and which showed protection in animal models.[44-45] The second approach started from the observation that clinical isolates belong to two types: those that produce a cytotoxin (VacA) and a cytotoxin-associated antigen (CagA) that are usually associated with the most severe forms of disease, such as peptic ulcer and gastric cancer, and those that do not produce these molecules and that are usually associated with milder forms of diseases, such as gastritis.[46] Genetic studies have shown that the difference between the two types of strains resides in 40 kilobases of DNA that are present only in type I bacteria that code for CagA and many other surface-associated proteins that are homologous to proteins found in secretion machineries of other bacteria and that confer virulence.[47] Candidate vaccine molecules have been selected among the proteins unique to the more pathogenic bacteria, following the rationale that

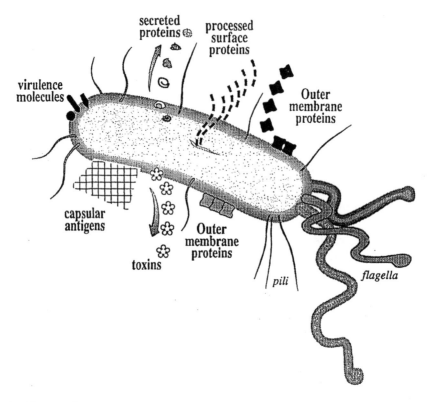

Figure 1.1. Bacterial antigens most frequently selected as components of vaccines. Note that these antigens are located at the surface of the bacteria or are secreted molecules.

the immune attack on these molecules should interfere with components essential for their virulence. Therefore, VacA, CagA, and other surface-associated molecules have been selected as vaccine candidates and have been shown to be protective in the animal model.[46]

In conclusion, candidate vaccine molecules should be preferentially selected among those that are unique to the pathogenic forms of bacteria. Figure 1.1 is a schematic representation of bacterial antigens that are most often used as vaccine components. These antigens, usually secreted or exposed at the cell surface, play a key role in virulence.

ANIMAL MODELS AND VACCINE DELIVERY

The first questions in the steps toward vaccine development are what is the mechanism of protection and which is the best way to induce it. A great help in elucidating the mechanism(s) of immunity is provided by the availability of an animal model. When a model is not available, the first research effort should be dedicated to its development. However, only rarely can the animal models fully mimic human disease; thus, we should keep in mind that although useful, animal models give only an indication of how to solve a problem and that the ultimate response will always come from testing the vaccine in humans. Chapter 2 contains further information on the use of animals in vaccine development.

Vaccines may target either a mucosal or a systemic immunity. Most of the successful vaccines are delivered by injection and target the systemic immunity. However, today, mucosal vaccines are becoming very popular, and a great effort is being made for their development. In theory, mucosal vaccines should be used for those pathogens that enter the body through the mucosal system (the portal of entry used by most pathogens). In practice, successful systemic vaccines have been developed against many pathogens that enter the host through the mucosae, and the development of mucosal vaccines has been restricted to those diseases for which systemic vaccination showed no effect.

WHAT TYPE(S) OF IMMUNE RESPONSES?

Until very recently, the approach to vaccine development has basically been empirical. Most of the vaccines used throughout the world have been generated by conventional microbiological techniques, which allowed the growth of the microorganisms *in vitro*, and then their attenuation by multiple passages in culture or their inactivation by physical and/or chemical procedures, which were also applied for the inactivation of microbial toxins. Except for the case of bacterial toxins, where the major pathogenic event was clear, such empirical approaches did not consider (technology at that time did not allow it) the complex interactions between microorganisms and their hosts, both in terms of the pathogenic mechanisms and in terms of immune responses elicited in the hosts. If the disease is considered as the result of the reequilibrium established between the pathogenic factors of the microbes on one side and the immune responses of the host on the other, it is evident that knowledge of these factors and responses would be critical in the design of new and/or improved vaccines, so as to move the equilibrium toward the establishment of powerful immunity before (preventive vaccines) or after (therapeutic vaccines) the encounter with the microorganism.

It is now clear that immune response in general and immunity in particular are complex and multifaceted phenomena which see the participation of different cell populations and soluble molecules that interact and regulate each other with positive or negative signals. The outcome of the immune response can be beneficial, as in the case of several viral diseases where viruses are eliminated and a state of immunity is established; neutral, as in the case of chronic infections (e.g., tuberculosis, *H. pylori* gastritis); or detrimental, as in the case of immune-mediated pathological events (e.g., CD8$^+$ cell-mediated liver cell damage in hepatitis B, and cerebral malaria). It is thus evident that, to be efficacious, vaccines must induce immune responses that are protective in both quantitative and qualitative terms, without triggering unwanted side effects.

Immune response is initiated when foreign organisms or antigens introduced into the body are taken up by professional cells (antigen-presenting cells) that enzymatically process them and reexpress peptide fragments in the context of the class I or class II molecules of the major histocompatibility complex (MHC), a phenomenon known as antigen presentation. Migration of these cells to draining lymph nodes will trigger specific activation of T cells that recognize the MHC-peptide complex through their antigen-specific receptor. The pathway of antigen processing, cytoplasmic or lysosomal, will dictate the class of MHC molecules involved in antigen presentation, class I or class II, respectively, and in turn the type of T cells, CD8[+] or CD4[+], respectively, that will be activated and engaged in proliferative phenomena.[48] These cells will then participate in effector functions against the microorganisms, either directly; for example, through cytolysis, production of cytokines such as IFN-γ by both CD8[+] and CD4[+] cells; or indirectly, for example, through the help provided by CD4[+] cells to B lymphocytes for the production of antigen-specific antibodies.

It is now clear that CD4[+] cells comprise functionally heterogenous subpopulations. In fact, on the basis of their ability to produce different patterns of cytokines, cloned murine and human CD4[+] T lymphocytes have been defined as T helper 1 (Th1) and Th2 cells.[49] Th1 cells produce IL-2, IFN-γ, lymphotoxin, etc.; mediate delayed-type hypersensitivity; and have some helper effects on B lymphocytes, e.g., in the production of IgG$_{2a}$ in mice. Conversely, Th2 cells produce IL-4, IL-5, IL-6, IL-10, IL-13, etc. and exhibit a strong helper effect in the production of several immunoglobulins, including IgE. Schematic representations of the ways to induce and measure the different types of CD4[+] immune responses are shown in Figures 1.2, 1.3, and 1.4.

Preferential expansion of one or the other CD4[+] T cell population can enormously influence the outcome of infections.[49] The best example known is that of the infection of mice with *Leishmania major*, intracellular protozoa infecting macrophages. During infection with *L. major*, susceptible BALB/c mice mount a polarized CD4[+] Th2-type response, whereas resistant mice mount a polarized CD4[+] Th1-type response.[50-51] Th1-type responses have been associated with protection in experimental murine models against some bacterial infections; e.g., with *Chlamydia*,[52] *B. pertussis*,[53] and *Listeria monocytogenes*.[54] On the other hand, protection against helminthic diseases has been reported as linked to induction of CD4[+] Th2-type responses.[55]

A wide variety of factors can intervene in influencing the polarization of the CD4[+] cell response toward one or the other functional phenotype. Genetic factors, the type of antigen-presenting cells, the amount of stimulating antigen, and the type of co-stimulatory molecules, such as B7, have all been involved.[49] However, if it is well accepted that IL-4 and IL-12 play a pivotal role in the differentiation of Th2 and Th1 cells, respectively, from a common Th0 precursor, little is still known about which cells preferentially and predominantly produce either cytokine, which will then trigger polarized differentiation and expansion of CD4[+] cell populations. CD3[+]CD4[+]NK1.1[+] cells exhibit a very early burst of IL-4 within a few hours after *in vivo* injection of anti-CD3 monoclonal antibody.[56] Following infection of mice with *L. major*, a similar early burst of IL-4 is observed in susceptible mouse strains, for which, however, CD3[+]CD4[+]NK1.1[+] lymphocytes are responsible.[57] On the other hand, macrophages and dendritic

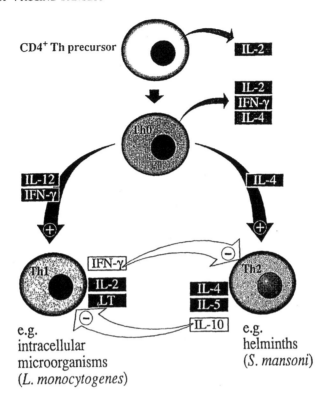

Figure 1.2. Schematic representation of the differentiation of CD4+ T lymphocytes into Th1 and Th2 cells.

cells may be the origin of the early production of IL-12, responsible for the differentiation of Th1-type cells either directly or through IFN-γ (e.g., from NK cells), as has been observed with some intracellular pathogens, such as *L. monocytogenes*.[54]

Detailed knowledge of the type of the effector mechanisms triggered by infection and the mechanisms of this triggering is critical if effective vaccines are to be designed with the aim of evoking or potentiating the immune response(s) that will confer strong and long-lasting immunity. This knowledge is also particularly important considering the fact that vaccine constructs can differently modulate the outcome of the immune response. For example, recombinant proteins and synthetic peptides essentially evoke CD4+ cell responses, as do the majority of soluble proteins;[48] thus, appropriate strategies are required for induction of cytolytic CD8+ cells, such as the use of viral or bacterial vectors,[17-19] particular corpusculate vehicles containing lipid tails,[58] and naked DNA immunization.[41] On the other hand, adjuvants can strongly influence the quality of the CD4+ cell subpopulation activated and, in turn, the quality of the antibody responses elicited.[58-59] For example, in mice antigen administration in aluminium salts (the only adjuvants admitted so far for human use) tends to favor the induction of Th2 cells, whereas Freund's adjuvants tend to favor Th1 cells. The

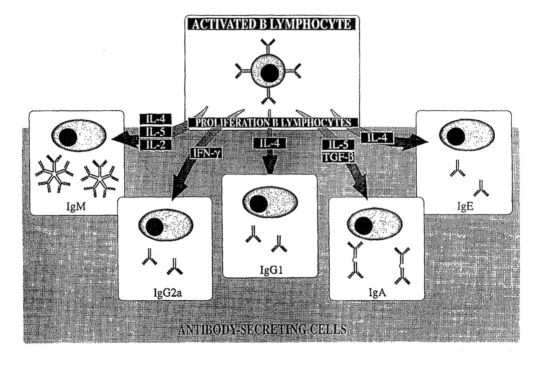

Figure 1.3. Cytokines produced by Th1 and Th2 cells play different roles in the signals required by B cells to produce different isotypes of antibodies.

routes of vaccine administration can also influence the quality of the immune response induced. Vaccination at the mucosal level, especially in conjunction with mucosally targeted adjuvants, such as *V. cholerae* and enterotoxigenic *Escherichia coli* toxins, strongly induce IgA responses, which can play a crucial role in the immunity against pathogens penetrating into the host through mucosal surfaces.[60]

In the development of new and/or improved vaccines, it is thus essential to know which kind of immune response is important to induce, and to select the most suitable strategies to induce it. The feasibility of the approach may limit the selection of vaccine targets.

REFERENCES

1. World Health Organization. The World Health Report 1996. Fighting disease, fostering development. Report of the Director-General. Geneva: WHO, 1996.
2. World Health Organization. World Health Organization Report 1993. *Investing in Health.* Washington: Oxford University Press, 1993.
3. National Institutes of Health. Accelerated Development of Vaccines 1994. The Jordan Report. Annual Report. Bethesda: NIH, 1996.
4. Centers for Disease Control and Prevention. 1996. *Morbidity and Mortality Weekly Report,* 45:805–809.

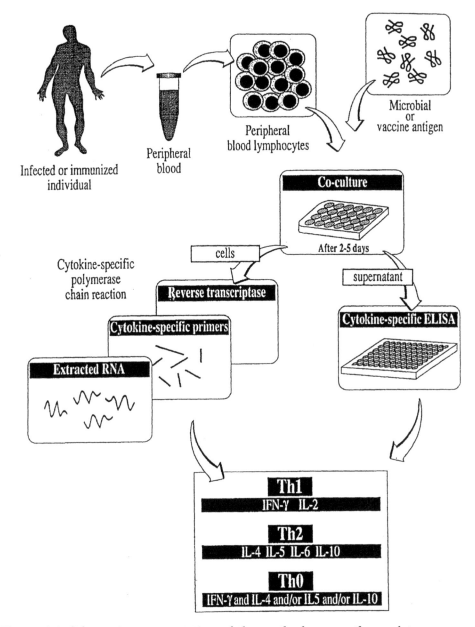

Figure 1.4. Schematic representation of the methods currently used to measure the activation of particular CD4+ T-cell subsets following natural infection or active immunization.

5. Harboe, M., P. Andersen, M.J. Colston, B. Gicquel, P.W.M. Hermans, J. Ivanyi, and S.H.E. Kaufmann. 1996. Vaccines against tuberculosis. *Vaccine.* 14:701–716.
6. Lowrie, D.B., R.E. Tascon, and C.L. Silva. 1995. Vaccination against tuberculosis. *Int. Arch. Allergy. Immunol.* 108:309–312.
7. Sabin, A.B. and L.R. Boulger. 1973. History of Sabin attentuated poliovirus live vaccine strains. *Biol. Stand.* 1:115.

8. Gluck, R. and M. Just. 1991. Vaccination against measles, mumps, and rubella, In: *Vaccines and Immunotherapy*, pp. 282–291, S.J. Cryz, Jr. (Ed.). Pergamon Press, New York.

9. Takahashi, M., T. Otsuka, Y. Okuno, Y. Asano, T. Yazaki, and S. Isomura. 1974. Live attenuated varicella vaccine used to prevent the spread of varicella in hospital. *Lancet.* 2:1288–1290.

10. Takahashi, M. 1986. Clinical overview of varicella vaccine: Development and early studies. *Pediatrics.* 78:736–741.

11. Germanier, R. and E. Furer. 1975. Isolation and characterization of galE mutant and Ty21a of Salmonella typhi: A candidate strain for a live oral typhoid vaccine. *J. Infect. Dis.* 141:553.

12. Everest, P., P. Griffiths, and G. Dougan. 1995. Live Salmonella vaccines as a route toward oral immunization. *Biologicals* 23:119–124.

13. Kaper, J.B., J.G. Morris, and M.M. Levine. 1995. Cholera. *Clin. Microbiol. Rev.* 8:48–86.

14. Levine, M.M., D. Herrington, G. Losonsky, B. Tall, J.B. Kaper, J. Ketley, C.O. Tacket, and S. Cryz. 1988. Safety, immunogenicity and efficacy of recombinant live oral cholera vaccines, CVD 103 and CVD 103-HgR. *Lancet* I:467–470.

15. Paoletti, E., J. Taylor, B. Meignier, C. Meric, and J. Tartaglia. 1995. Highly attenuated poxvirus vectors: NYVAC, ALVAC and TROVAC. *Dev. Biol. Stand.* 85:159–163.

16. Tartaglia, J., M.E. Perkus, J. Taylor, E.K. Norton, J.-C. Audonnet, W.I. Cox, S.W. Davis, J. Vanderhoeven, B. Meignier, M. Riviere, B. Languet, and E. Paoletti. 1992. NYVAC: A highly attentuated strain of vaccinia virus. *Virology.* 188:217–232.

17. Esposito, J.J. and F.A. Murphy. 1989. Infectious recombinant vectored virus vaccines, In: *Vaccine Biotechnology*, pp. 196–247, Volume 33. Advances in veterinary science and comparative medicine, Bittle J.L. and F.A. Murphy, (Eds.). Academic Press, Inc., San Diego.

18. Norrby, E. 1989. Modern approaches to live virus vaccines, In: *Vaccine Biotechnology*, pp. 249–270, Volume 33. Advances in veterinary science and comparative medicine, Bittle J.L. and F.A. Murphy, (Eds.). Academic Press, Inc., San Diego.

19. Dougan, G., L. Smith, and F. Heffron. 1989. Live bacterial vaccines and their application as carriers for foreign antigens, In: *Vaccine Biotechnology*, pp. 271–300, Volume 33, Bittle, J.L. and F.A. Murphy (Eds.). Academic Press, Inc., San Diego.

20. Manclark, C.R. and J.L. Cowell. 1984. Pertussis, In: *Bacterial Vaccines*, pp. 69–106. Germanier, R. (Ed.), Academic Press, Inc., Orlando.

21. Salk, J.E. 1960. Persistence of immunity after administration of formalin treated poliovirus vaccine. *Lancet* 2:715.

22. Wood, J.M. 1991. Influenza vaccines, In: *Vaccines and Immunotherapy*, pp. 271–281, Cryz, S.J., Jr. (Ed.). Pergamon Press, New York.

23. Celis, E., C. Rupprecht, and S.A. Plotkin. 1990. New and improved vaccines against rabies, In: *New Generation Vaccines*, pp. 419–438, Woodrow, G.C. and M.M. Levine (Eds.), Marcel Dekker Inc., New York.

24. Stephenson, J.R., J.M. Lee, L.M. Easterbrook, A.V. Timofeev, and L.B. Elbert. 1995. Rapid vaccination protocols for commercial vaccines against tick-borne encephalitis. *Vaccine* 13:743–746.

25. Rappuoli, R. 1990. New and improved vaccines against diphtheria and tetanus, In: *New Generation Vaccines*, pp. 251–268, In Woodrow, G.C. and M.M. Levine (Eds.), Marcel Dekker, Inc., New York.

26. Jennings, H.J. 1983. Capsular polysaccharides as human vaccines. *Adv. Carbohydr. Chem. Biochem.* 41:155.

27. Gotschlich, E.C. 1984. Meningococcal meningitis, In: *Bacterial Vaccines*, pp. 237–255. Germanier, R. (Ed.), Academic Press, Inc., Orlando.

28. Robbins, J.B., R. Schneerson, and M. Pittman. 1984. Haemophilus influenzae type B Infections, In: *Bacterial Vaccines*, pp. 290–316, Germanier, R. (Ed.), Academic Press, Inc., Orlando.

29. Costantino, P., S. Viti, A. Podda, M.A. Velmonte, L. Nencioni, and R. Rappuoli. 1992. Development and phase I clinical testing of a conjugate vaccine against meningococcus A and C. *Vaccine* 10:691–698.

30. Lieberman, J.M., S.S. Chiu, V.K. Wong, S. Partridge, S.J. Chang, C.Y. Chiu, L.L. Gheesling, G.M. Carlone, and J.I. Ward. 1996. Safety and immunogenicity of a serogroups A/C Neisseria meningitidis oligosaccharide-protein conjugate vaccine in young children: A randomized controlled trial. *JAMA* 275:1499–1503.

31. Szu, S.C., D.N. Taylor, A.C. Trofa, J.D. Clements, J. Shiloach, J.C. Sadoff, D.A. Bryla, and J.B. Robbins. 1994. Laboratory and preliminary clinical characterization of Vi capsular polysaccharide-protein conjugate vaccines. *Infect. Immun.* 62:4440–4444.

32. Wessels, M.R., L.C. Paoletti, A.K. Rodewald, F. Michon, J. Difabio, H.J. Jennings, and D.L. Kasper. 1993. Stimulation of protective antibodies against type-Ia and type-Ib group-B streptococci by a type-Ia polysaccharide- tetanus toxoid conjugate vaccine. *Infect. Immun.* 61:4760–4766.

33. Ellis, R.W. 1990. New and improved vaccines against hepatitis B, In: *New Generation Vaccines*, pp. 439–447, Woodrow, G.C. and M.M. Levine (Eds.), Marcel Dekker, Inc., New York.

34. Brennan, M.J., D.L. Burns, B.D. Meade, R.D. Shahin, and C.R. Manclark. 1992. Recent advances in the development of pertussis vaccines, In: *Vaccines. New Approaches to Immunological Problems*. Ellis, R.W. (Ed.), Butterworth-Heinemann, Boston, London.

35. Decker, M.D., K.M. Edwards, M.C. Steinhoff, M.B. Rennels, M.E. Pichichero, J.A. Englund, E.L. Anderson, M.A. Deloria, and G.F. Reed. 1995. Comparison of 13 acellular pertussis vaccines: Adverse reactions. *Pediatrics* 96:557–566.

36. Edwards, K.E., B.D. Meade, M.D. Decker, G.F. Reed, M.B. Rennels, M.C. Steinhoff, E.L. Anderson, J.A. Englund, M.E. Pichichero, M.A. Deloria, and A. Deforest. 1995. Comparison of thirteen acellular pertussis vaccines: Overview and serologic response. *Pediatrics* 96:548–557.

37. Moxon, E.R. and R. Rappuoli. 1990. Haemophilus influenzae infections and whooping cough. *Lancet* 335:1324–1329.

38. Greco, D., S. Salmaso, P. Mastrantonio, M. Giuliano, A.E. Tozzi, A. Anemona, M.L.C.D. Atti, A. Giammanco, P. Panei, W.C. Blackwelder, D.L. Klein, S.G.F. Wassilak, P. Stefanelli, M. Bottone, T. Sofia, S. Luzi, G. Bellomi, F. Cobianchi,

G. Canganella, F. Meduri, G. Scuderi, A. Chiarini, M. Maggio, S. Taormina, M. Genovese, A. Moiraghi, A. Barale, S. Ditommaso, S. Malaspina, E. Vasile, P. Ferraro, P. Dallago, L. Demarzi, L. Robino, E. Giraldo, N. Coppola, P. Materassi, G. T. Castellani, F. Basso, S. Barbuti, M. Quarto, P. Lopalco, P. Dorazio, and A. Sanguedolce. 1996. A controlled trial of two acellular vaccines and one whole-cell vaccine against pertussis. *N. Engl. J. Med.* 334:341–348.

39. Pizza, M., A. Covacci, A. Bartoloni, M. Perugini, L. Nencioni, M.T. de Magistris, L. Villa, D. Nucci, R. Manetti, M. Bugnoli, F. Giovannoni, R. Olivieri, J.T. Barbieri, H. Sato, and R. Rappuoli. 1989. Mutants of pertussis toxin suitable for vaccine development. *Science* 246:497–500.

40. Rappuoli, R., G. Douce, G. Dougan, and M. Pizza. 1995. Genetic detoxification of bacterial toxins: A new approach to vaccine development. *Int. Arch. Allergy Immunol.* 108:327–333.

41. Ulmer, J.B., D.L. Montgomery, J.J. Donnelly, and M.A. Liu. 1996. DNA vaccines, In: *Vaccine Protocols*, pp. 289–300, Robinson, A., G.H. Farrar, and C.N. Wiblin (Eds.), Humana Press, Totowa, NJ.

42. Weiss, A. and E.L. Hewlett. 1986. Virulence factors of *Bordetella pertussis. Annu. Rev. Microbiol.* 40:661–686.

43. Weiss, A.A., E.L. Hewlett, G.A. Myers, and S. Falkow. 1984. Pertussis toxin and extracytoplasmic adenylate cyclase as virulence factors of *Bordetella pertussis. J. Infect. Dis.* 150:219–222.

44. Ferrero, R.L., J.M. Thiberge, M. Huerre, and A. Labigne. 1994. Recombinant antigens prepared from the urease subunits of *Helicobacter* spp: Evidence of protection in a mouse model of gastric infection. *Infect. Immun.* 62:4981–4989.

45. Michetti, P., I. Corthesy-Theulaz, C. Davin, R. Haas, A.C. Vaney, M. Heitz, J. Bille, J.P. Kraehenbuhl, E. Saraga, and A.L. Blum. 1994. Immunization of BALB/c mice against *Helicobacter felis* infection with *Helicobacter pylori* urease. *Gastroenterology* 107:1002–1011.

46. Marchetti, M., B. Aricò, D. Burroni, N. Figura, R. Rappuoli, and P. Ghiara. 1995. Development of a mouse model of *Helicobacter pylori* infection that mimics human disease. *Science* 267:1655–1658.

47. Censini, S., C. Lange, Z. Xiang, J.E. Crabtree, P. Ghiara, M. Borodovsky, R. Rappuoli, and A. Covacci. 1996. Cag, a pathogenicity island of *Helicobacter pylori*, encodes type I-specific and disease-associated virulence factors. *Proc. Natl. Acad. Sci. USA* 93:14648–14653.

48. Germain, R.N. 1994. MHC-dependent antigen processing and peptide presentation: providing ligands for lymphocyte activation. *Cell* 76:287–299.

49. Mosmann, T.R. and S. Sad. 1996. The expanding universe of T-cell subsets: Th1, Th2 and more. *Immunol. Today* 17:138–146.

50. Milon, G., G. Del Giudice, and J.A. Louis. 1995. Immunobiology of experimental cutaneous leishmaniasis. *Parasitol. Today* 11:244–247.

51. Reiner, S.L. and R.M. Locksley. 1995. The regulation of immunity to Leishmania major. *Annu. Rev. Immunol.* 13:151–177.

52. Rank, R.G., K.H. Ramsey, E.A. Pack, and D.M. Williams. 1992. Effect of gamma interferon on resolution of murine chlamydial genital infection. *Infect. Immun.* 60:4427–4429.

53. Mills, K.H.G., A. Barnard, J. Watkins, and K. Redhead. 1993. Cell-mediated immunity to *Bordetella pertussis*: role of Th1 cells in bacterial clearance in a murine respiratory infection model. *Infect. Immun.* 61:399–410.
54. Hsieh, C.S., S.E. Macatonia, C.S. Tripp, S.F. Wolf, A. O'Garra, and K.M. Murphy. 1993. Development of Th1 CD4+ T cells through IL-12 produced by Listeria-induced macrophages. *Science* 260:547–549.
55. Oswald, I.P., P. Caspar, D. Jankovic, T.A. Wynn, E.J. Pearce, and A. Sher. 1994. IL-12 inhibits Th2 cytokine responses induced by eggs of *Schistosoma mansoni*. *J. Immunol.* 153:1707–1713.
56. Yoshimoto, T. and W.E. Paul. 1994. CD4+, NK1.1+ T cells promptly produce interleukin 4 in response to *in vivo* challenge with anti-CD3. *J. Exp. Med.* 179:1285–1295.
57. Launois, P., T. Ohteki, K. Swihart, H.R. MacDonald, and J.A. Louis. 1995. In susceptible mice *Leishmania major* induce very rapid interleukin 4 production by CD4+ cells which are NK1.1-. *Eur. J. Immunol.* 25:3298–3307.
58. Morein, B., K. Lövgren-Bengtsson, and J. Cox. 1996. Modern adjuvants. Functional aspects, In: *Concepts in Vaccine Development*, pp. 243–263, Kaufmann, S.H.E. (Ed.), Walter de Gruyter, Berlin, New York.
59. Del Giudice, G. 1992. New carrier and adjuvants in the development of vaccines. *Curr. Opin. Immunol.* 4:454–459.
60. Douce, G., C. Turcotte, I. Cropley, M. Roberts, M. Pizza, M. Domenighini, R. Rappuoli, and G. Dougan. 1995. Mutants of *Escherichia coli* heat-labile toxin lacking ADP-ribosyl-transferase activity act as nontoxic, mucosal adjuvants. *Proc. Natl. Acad. Sci. USA* 92:1644–1648.

2 Use of Animals for Vaccine Development

Andrew B. Onderdonk and Ronald C. Kennedy

GENERAL CONSIDERATIONS

Animals have been a source of research interest since ancient times. In addition to serving as "beasts of burden" and as a food supply for early civilizations, animals clearly provided humans with early lessons in anatomy and physiology. In the modern era, the earliest attempts to use animals for vaccine development date to Pasteur and Koch.[1] Chickens were used in one of the first successful attempts to develop a vaccine against the bacterial disease fowl cholera, caused by *Pasteurella multocida*. In recent decades as the development and testing of vaccines has become more sophisticated, our need for legitimate animal models of human infections has increased.[2-7] During the developmental process for any biologic product requiring FDA approval, animals must be used for safety and efficacy testing before human clinical trials.[7] For vaccines, there is the added issue of documenting both immunogenicity and protection against a specific disease process that occurs in humans. Animal models for human infections are, therefore, an important early component for vaccine development.

An animal model is an infectious process that mimics infection in humans. Animal models include both naturally occurring infections of animals and infections induced by specific microbiologic challenge.[8,9] The purpose of this chapter is to familiarize the reader with the nature of such models in animals ranging from laboratory rodents to nonhuman primates. Issues regarding the selection of an appropriate animal species and how animals should be housed and cared for are included because they are essential to the successful use of animal models in research.

Animal Models and Animal Test Systems

An important first step in the selection of an animal species for use as part of vaccine development is to understand the differences between animal models

Table 2.1. Attributes of Animal Models and Animal Test Systems

	Animal Models	Animal Test System
Species used	Many	Rodents, rabbits
Method of induction	Naturally occurring or induced	Induced
Endpoints for evaluation	Same as human infection	Mortality or other easily measured parameter
Purpose	Simulate human disease	Test toxicity or potential therapy

and animal test systems (Table 2.1). An animal model develops a disease that closely mimics its counterpart in humans. For a particular animal species to be considered for use in a model, the organism of interest must be able to grow and cause disease in the target species. In order for a model to be successful, the endpoints for evaluation must be the same as those used for evaluating the human disease process.[9,10] Endpoints often include mortality, positive blood and organ cultures, response to therapy, physiologic parameters such as white blood cell counts or serum chemistry values, and immunologic parameters such as the development of a specific antibody. Animal models can be either naturally occurring; for example, poliovirus, *Shigella* species or *Mycobacterium tuberculosis* infections in nonhuman primates, rabies infection in a variety of animal species, and streptococcal mastitis in cows; or induced; for example, the rat model for intraabdominal sepsis,[11,12] *Clostridium difficile* colitis in hamsters,[13–16] and HIV infection in nonhuman primates.[17,18] In each case, the etiologic agent of the disease is the same as that causing the human disease, and the endpoints for evaluation are similar to those in humans. While not all animal models are absolutely faithful to the events noted in humans, such as the neonatal mouse model for group B streptococcal infection,[19] variance from the human equivalent is usually due to physiologic and/or anatomic differences in the model species.

Animal test systems use an animal species that is "susceptible" to challenge with an infectious agent. In an animal test system, the route of challenge, the progression of the disease, the endpoints for evaluation, and the host response to infection do not necessarily mimic the disease process in humans. Animal test systems are often used as a method for screening in rodents potentially toxic or therapeutic compounds that may subsequently be tested in an authentic animal model before human clinical trials. Because animal test systems are a relatively inexpensive method for evaluating biologic activity, they are widely used in the research community. However, such test systems should not be confused with, or called, animal models. Examples of animal test systems include *Staphylococcus aureus* or *Escherichia coli* peritonitis in mice for evaluating antibiotic efficacy, *Salmonella enteritidis* infection in mice, and *Legionella* infections in guinea pigs. The endpoint for most animal test systems used for infectious agents is mortality, although other endpoints may also be used. The primary use of an animal test system is to provide an inexpensive replication of infection with endpoints that are easy to measure.[2,3,6]

Use of Animal Models During Vaccine Development

Animal models are used during the development of vaccines in many different ways, perhaps the most important being to determine whether an antigen isolated from a particular infectious agent provokes protective immunity against challenge with the specific agent. Animals are often necessary to completely characterize the pathogenic features and virulence mechanisms for the bacterial or viral infection against which the vaccine is directed. Animals are also important for determining the host response to various permutations and combinations of antigens, adjuvants, and immunization schedules. Studies of immunogenicity in an animal model often determine the actual vaccine candidate(s) to be tested for safety and efficacy, and to be used in human trials.

Following initial trials in an animal model, potential vaccines must be proved safe for human use before human testing. Animal testing in this case generally includes more than one animal species and need not be performed in an animal model. Since the primary safety issue is toxicity of a vaccine formulation and host physiologic response to a foreign antigen, tests are performed in both small animals, such as rodents, guinea pigs, or rabbits, and larger animal species, including dogs and nonhuman primates. Efficacy studies, on the other hand, require that a susceptible animal species be vaccinated, the immune response documented, and protection against challenge with the bacterial or viral agent proven. This type of testing is most often performed in an animal model of the human disease.

The final use of animals in vaccine development is to determine whether scale-up procedures for producing and storing vaccines in commercial quantities are adequate. Lot testing of vaccines to certify protective efficacy and safety continues to often be a requirement for release of vaccines for human immunization.[4-7,20-26]

Selecting an Animal Species for Use

The starting point for developing an animal model of a human infection is the scientific literature. Literally hundreds of articles detail experimental infections in a wide variety of animal species. Investigators involved in vaccine development should review this literature, paying careful attention to the attributes of the described "models." Basic questions to be answered should include the similarity of the process in animals to that in humans, the reproducibility of the model system, and the endpoints that are actually being measured. Additional considerations should encompass the availability and cost of animals, ease of use, and similarity between animal and human host immune response to the infectious challenge. For rodent models, investigators may wish to consider whether it is better to use a genetically defined, inbred strain of mouse, or an outbred strain. While the reproducibility of the inbred strain has many advantages, particularly for immunologic studies, humans are not an inbred strain and therefore do not necessarily respond in an identical manner to the same infectious insult. Outbred animals may be more difficult to use for detailed immunologic studies, but they tend to yield a more robust model system demonstrating a spectrum of disease.

If no animal model for a particular disease process exists, the investigator's task is more difficult. First, an animal species susceptible to the organism of interest must be identified, a difficult task with agents such as *Mycobacterium leprae* (armadillo) and *Treponema pallidum* (rabbit testes), which are fastidious and slow-growing pathogens. Viral agents are even more difficult to work with, since the tissue tropisms of many viral agents are well-known and often limited to a particular species. For bacterial pathogens, rodents, particularly mice, can usually be used to simulate at least some aspect of the human disease. With the advent of transgenic and knockout mouse strains, many potentially problematic immunologic issues can be circumvented by the use of strains that either contain a specific human gene or lack a specific mouse gene. The tests for virtually all animal models of human infections are the same: how well does the experimental disease simulate human experience, is the disease reproducible, and are the endpoints for evaluation the same as those in humans? Assessment of the utility of any species should also include considerations of cost and availability of animals. If the species of interest is either exotic or endangered, ethical considerations should prevail. A list of some commonly employed animal model systems, the species used, and the method for induction is given in Table 2.2.

Nonhuman Primates in Vaccine Development

The use of nonhuman primates in biomedical research related to vaccine development had its origin in Pasteur's work on rabies virus. Other investigators used nonhuman primates in studies with smallpox and vaccinia in the late 1800s. Perhaps the major focus on the utility of nonhuman primates in biomedical research came with the Nobel prize work of Landsteiner and Popper,[27,28] who employed rhesus monkeys, baboons, and chimpanzees in research that resulted in the isolation of the poliovirus. The unique susceptibility of nonhuman primates to poliovirus infection established their importance in future biomedical research efforts.

Nonhuman primate species were used in the development and testing of inactivated or partially inactivated polio vaccines. In the early 1930s, monkeys were inoculated with throat washings from patients with poliomyelitis and became infected.[29] Subsequent work in nonhuman primates demonstrated that poliomyelitis was an enteric infection. Although the Nobel prize work by Enders, who developed a means of *in vitro* propagation of poliovirus in human tissue culture, was a major scientific advance,[30] nonhuman primates still played a major role in polio vaccine development.

The formalin-inactivated polio vaccine that was grown in monkey kidney cells was employed in the first extensive and largely successful field trials.[31]

Table 2.2. Commonly Employed Animal Models for Human Infections

Disease	Species	Method of Induction	Endpoints
Intraabdominal sepsis	Rat	Surgical implantation of intestinal contents	Mortality or abscess
Endocarditis	Rabbit	Catheterization and bacterial challenge	Vegetations
Otitis media	Chinchilla	Bacterial challenge	Histopathology, culture
Renal abscess	Rat	Bacterial challenge via parenteral route	Abscess, culture
Antibiotic-associated colitis	Hamster	Administration of antibiotics	Mortality, histopathology, toxin assays
Polio	Baboon	Challenge	Serology, viral cultures
Hepatitis A,B,C	Chimpanzee	Challenge	Serology, LFT histopathology
RSV	Infant chimpanzee	Intranasal challenge	Varied
Pertussis	Marmoset	Intranasal challenge	Clinical symptoms, culture
Group B *Streptococcus*	Mice, macaques	Challenge	Mortality, culture

However, enthusiasm over this achievement was dampened by reports of improper inactivation of the vaccine, which resulted in a number of cases of polio in human vaccine recipients.[32] Development of an attenuated poliovirus for vaccine usage was dependent on the use of monkeys and chimpanzees.[25] The work to develop an oral polio vaccine reportedly employed 9,000 monkeys and 150 chimpanzees.[25]

ANIMAL CARE CONSIDERATIONS

The care and housing of laboratory animals is an important part of any animal research study. For rodent populations, many investigators choose to use virus antibody-free (VAF) animals because of the potential for immunosuppression caused by many murine viruses, such as Sendai, mouse hepatitis virus (MHV), and minute virus of mice (MVM). Specific pathogen-free (SPF) animals have been tested to be free of specific viral, mycoplasma, and bacterial pathogens along with endo- and ectoparasites but may or may not be VAF animals. Such animals are raised in barrier facilities by the vendor and are more expensive than animals that are raised in conventional housing units. The advantage of using VAF and/or SPF animals is that they tend to provide uniform responses to infectious challenge and to immunization. In addition, interference with challenge studies by animals co-infected with agents such as mycoplasma or Sendai virus can be avoided. Both transgenic and knockout animals are currently in use as animal models and often require special barrier containment because of their susceptibility to extraneous infections. In some instances, germfree or gnotobiotic animals are also used for bacterial or viral challenge studies. Each type of animal has unique requirements for housing and general care. The American Association for Laboratory Animal Care (AALAC) International guidelines provide for specific minimal housing and surveillance procedures for each species within an animal housing facility. These guidelines assist the investigator in determining the most humane care for a particular species.[33,34]

For studies that involve infectious challenge or work with an infectious agent that is a human pathogen, biosafety facilities may be needed. Such facilities raise the cost of animal housing. The four designated Animal Biosafety Levels (ABSL) correspond very closely to biosafety levels for handling infectious hazards within research and clinical laboratories. These four categories provide increasing levels of protection to personnel and the environment. The ABSL are the minimal recommended standards for activities involving laboratory animals potentially or known to be infected either naturally or experimentally. With some infectious agents, biosafety level 3 and 4 facilities may be required to house and maintain both rodents and nonhuman primates undergoing studies. Because there are few such facilities and those available are expensive to use, the investigator should be familiar with what is available before attempting to begin studies. ABSL facility descriptions, standards, and special practices are detailed in "Biosafety in Microbiological and Biomedical Laboratories."[34,35] The ABSL 1 involves work with viable microorganisms not known to cause disease in healthy adult humans, whereas ABSL 4 involves dangerous and exotic agents that pose a high individual risk of life-threatening disease, such as the hemorrhagic fever viruses. ABSL 2 practices, containment equipment, and facilities should be observed in the care and use of nonhuman primates. Higher levels of biosafety containment should be considered, with the emphasis on prevention and control of the spread of etiologic agents, when animals are undergoing experimental infection studies with human pathogens. Excellent reviews of this subject can be found elsewhere.[34,35]

USE OF RODENTS AS ANIMAL MODELS

Mice and rats are often the species of choice for development of animal model systems because of their relatively small size and low cost. Although strains of mice and rats are less expensive than larger animal species, costs of $6 to $25 or more per animal are not unusual for VAF and/or SPF inbred mouse and rat strains. Housing costs for rats and mice have increased dramatically in conjunction with the more stringent housing and care requirements mandated by the federal government. Nevertheless, these two species continue to be among the first considered by investigators when an animal model is required.

It is desirable to use an animal model that results from a naturally occurring infectious process, if available. However, such models occur rarely among rodent populations due to the substantial differences in anatomy, physiology, and endogenous microflora. Although there are naturally occurring processes in rodents similar to those in humans, such as the recent reports of inflammatory bowel diseases in various knockout mouse strains, these are the exception rather than the rule. The microbial species that cause infections in rodents are often different from those isolated from human populations. Microorganisms, such as *Borrelia burgdorferi*, can colonize deer mice naturally without causing apparent disease. Other agents, such as *Bordetella pertussis*, *Mycobacterium tuberculosis*, and *Listeria monocytogenes* do not colonize rodents in a natural setting, while species that are not common human pathogens, such as *Bordetella bronchiseptica*, *Staphylococcus intermedius*, and *Citrobacter freundii* can colonize a variety of animal species. The phenomenon of species specificity is particularly applicable to viral agents, which usually have a well-defined host range. Although mice and rats have their own indigenous viral agents, viruses such as Herpes simplex and Epstein Barr virus do not normally infect rodents, while agents such as lymphocyticchoriomeningitis (LCM) virus can be carried by and cause disease in hamsters. As a result of the lack of naturally occurring infections with human pathogens, most rodent models result from an induced infectious process.

Infectious Challenge

The development and use of an animal model for human infection is a complex process starting with the identification of an appropriate infectious insult. One such model is the rat model for intraabdominal sepsis.[11,12] First devised in 1974, it is an induced model simulating the biphasic disease process resulting from the contamination of the peritoneal cavity with the contents of the large intestine. The endpoints for untreated disease are early mortality associated with septicemia, followed by abscess development in surviving animals. The induction of intraabdominal sepsis relies on the surgical implantation of intestinal contents from meat-fed rats. The inoculum for this model proved to be the key to the successful simulation of the disease process in humans. When the model was first developed, little was known about the microorganisms responsible for the infectious process in humans beyond the fact that facultative gram-negative rods, such as *E. coli*, were commonly isolated from patients with accidental peritoneal soilage. Early experimentation with intestinal contents from conventional, grain-fed rats showed variable results for both endpoints. Microbiologic studies revealed that grain-fed rats harbored an intestinal microflora quite different from that of humans. Changing the rats' diet to lean ground beef produced a microflora similar to that in humans. When this inoculum was implanted into recipient animals, the resulting reproducible disease resembled that described for untreated intraabdominal infections in humans. During the development process it was determined that very subtle differences in the inoculum could alter the experimental disease process substantially. *E. coli* or *B. fragilis* strains of human, rather than rat, origin were shown to produce profoundly different effects when implanted into other animals. Titration of inocula containing *E. coli* from various sources indicated that the number of viable organisms required to produce significant mortality varied considerably from strain to strain. For most induced models of human infection, the preparation of the inoculum is one of the most important factors in the reproducibility of the model system. In the case of the rat model described above, a polymicrobic inoculum containing both obligate anaerobes and facultative species was used; however, even infections induced in rodents with a single organism such as group B *Streptococcus, Streptococcus pneumoniae, Helicobacter pylori, S. aureus* or *E. coli* require care in the preparation and administration of the infectious challenge.

Choice of strains to use for challenge is an important consideration. Generally, a strain that has been freshly isolated from the human disease and that has been passed in the target animal species is the starting point for animal infection models. Repeated passage of strains in the laboratory is not desirable since strain characteristics, such as encapsulation, extracellular enzyme, and toxin production, and surface characteristics of bacteria may change. Many investigators choose a strain, produce substantial amounts of first- passage material, and then preserve aliquots of the inoculum for future animal challenge. For frozen stock cultures, assays of viable cell density can determine whether the number of colony-forming units has decreased during storage and will ensure that animals receive a uniform infectious challenge. Viral agents grown in tissue culture can be preserved indefinitely in the frozen or lyophilized state. Assays for plaque-forming units, or infectious-dose (ID) units and quantitative DNA or RNA assays are useful as methods for adjusting inoculum dosage for viral agents.

Immunologic Considerations

The laboratory mouse has received the most attention as a model for immunologic studies because of its low cost, the ease in handling, and the availability of a variety of inbred strains. Until recently, rats were not available as inbred strains and therefore have not been studied as thoroughly as inbred mouse strains. Although a complete description of the immune system of the mouse is well beyond the scope of this chapter, several excellent references exist.[36] Mice have both humoral and cellular immune responses similar to those of humans. In addition, monoclonal antibodies directed at cell surface markers allow for specific immune cells, including T and B lymphocytes and the mouse equivalent of CD4 and CD8 antigens, to be targeted easily. The thorough study of the genetics of the mouse immune system has resulted in two new approaches to animal models based on genetic manipulation in transgenic and knockout mice (see below).

There are also several naturally occurring mouse strains with specific immune deficiencies. One of the first to be characterized, in 1966, was the hairless or "nude" mouse, which lacks a thymus and therefore has no functional T-lymphocyte populations. Because these mice can be reconstituted with specific T-cell subsets from congenic littermates, they have been enormously helpful for the study of cell- mediated immune responses as they relate to vaccine development. A second mouse strain, the SCID (severe combined immunodeficiency) mouse described in 1980, lacks both T and B cell immune functions. Both mouse strains require specialized housing due to their lack of immune competence. The nude mouse is very sensitive to infections with organisms such as *S. aureus* and *Pseudomonas* sp., while SCID mice are susceptible to a wide array of environmental organisms commonly present as normal microbiota of humans and other animals. Despite these drawbacks, both strains are quite useful for establishing an infectious process not easily produced in immunologically intact rodents, such as HIV encephalitis in the SCID mouse.[37] Immune-deficient strains also provide useful insights into immune mechanisms via passive protection experiments with hyperimmune globulin or by reconstitution with specific cell types.

A natural extension of our understanding of the immunology and genetics of nude and SCID mice led to the development of the transgenic animal in which specific DNA material not found in the mouse can be incorporated and phenotypically expressed in littermates homozygous for the particular characteristic. Thus, human genes, such as cell surface receptors for viral agents, human tumor cell markers, and specific immune functions can be expressed in a mouse system using an inbred genetic background. Inbred mouse strains that have been "gene targeted" by DNA injections, retrovirus infection, and/or homologous recombinations allow for chimeric animals to be cross-bred in order to turn off or knockout phenotypic expression related to a specific gene(s). Gene targeting permits a lack of phenotypic expression or, in the case of suppressor factors, the uniform expression of particular characteristics. Dozens of different knockout mouse strains are available, many commercially. Some of these strains are particularly useful because they are deficient in a characteristic important to disease development or host response to infection. For example, a mouse strain may lack MHC class I molecule expression, be incapable of processing antigens properly, or be deficient in differentiated immune cell function. Use of such strains in models for disease has added much new information regarding fundamental host response, although their use in genuine models for vaccine trials has not yet been attempted.

Many practical considerations should be kept in mind when using rodents as animal model systems. Remarkably, simple considerations, such as the type of adjuvant to be used with an experimental antigen or the immunization schedule, can cause the most problems for subsequent vaccine development. Adjuvants such as Freund's incomplete adjuvant, while highly potent in rodents with a wide array of antigens, cannot be used in humans. Evaluation of the immune response to an antigen that has been introduced with this adjuvant may or may not be relevant to the human response. If an adjuvant is going to be used for testing purposes, it is desirable to use one that can also be employed in larger animals including humans, thereby avoiding the problem of having to retest because of changes in formulation of the vaccine. It should also be pointed out that most polyclonal and monoclonal antibody reagents directed at cell surface markers or cytokines in humans do not react with the comparable markers in mice or rats.

Another consideration when dealing with rodent models is the basic reproductive cycle of rodents compared with that of humans. Mice have a gestation period of 22 days. Vaccines that might be directed at a mother during pregnancy in order to passively protect offspring, such as the conjugate vaccine being developed for group B *Streptococcus*, must take into account the shorter gestation period of the mouse and the timing of vaccination in order to measure passive protection in neonates.[38] Issues such as transplacental transfer of antibody, and how and when to challenge neonates are also important in this induced model.[19,39] Recent descriptions of the neonatal mouse model for GBS and the results of vaccination provide insight into technical issues involved with this particular vaccination strategy.

NONHUMAN PRIMATE MODELS TO EVALUATE VACCINE IMMUNOGENICITY

In preclinical studies that will evaluate the safety and immunogenicity of putative vaccine candidates, species most closely related to humans should be considered. Phylogenetically, the great apes (chimpanzees, orangutans, gorillas, and gibbons) are most closely related to humans. However, the great apes, which diverged from humans over 5 million years ago, are endangered or threatened species, which may limit their utility in preclinical studies. In addition, cost considerations for using great apes in biomedical research are another serious limitation. The Old World monkeys (macaques, baboons, mandrills, and mangabeys) diverged from humans over 15 million years ago. A number of the Old World monkey species including rhesus, cynomolgus, and African green monkeys, have been employed to evaluate vaccine safety and immunogenicity. New World monkeys (aotus, owl, and cebus monkeys and marmosets) are the most phylogenetically divergent from humans, yet they have also been used to develop nonhuman primate models for a number of human infectious diseases that have applications to vaccine-related research.

The phylogenetic relationships among primates as they relate to morphological synapomorphies have been addressed in detail. New molecular and biochemical approaches by comparison of DNA sequence and structure-function relationships continue to provide new information and conflicts. Chimpanzees (*Pan troglodytes*), orangutans (*Pongo pygmaeus*), gorillas (*Gorilla gorilla*), and gibbons (*Hylobates lars*) are most closely related to *Homo* species among the hominoid primates.[40] The Old World monkeys comprised of the cercopithecine tribe, Papionini are next in evolutionary distance.[41,42] This includes six living genera: drills and mandrills (*Mandrillus*); common or savannah baboons (*Papio*); gelada baboons (*Theropithecus*); mangabeys (*Cercocebus* and *Lopocebus*); and macaques (*Macaca*). The Papionins radiated during the late Miocene to Plio-Pleistocene period and are considered a sister group to the Hominoidea.

Regional Primate Research Centers Program

The increased applications of nonhuman primates to scientific research during the twentieth century resulted in the establishment of Primate Centers. The first primate breeding station was established in the former USSR in 1923. In 1930, an accomplished comparative psychologist, Robert Yerkes, established the Primate Laboratory of the Yale Institute of Psychobiology at Orange Park, Florida.[43,44]

The process that resulted in the establishment of the NIH Regional Primate Research Centers Program (RPRCP) has been described in detail.[45,46] Its origins date back to 1947 and 1949 when the NIH was unsuccessful in procuring a supply of chimpanzees for researchers in the United States. In 1961, Congress appropriated an additional $7 million and grants were awarded to establish the Washington RPRC at the University of Washington in Seattle; the Wisconsin RPRC at the University of Wisconsin in Madison; the Yerkes RPRC in association with Emory University in Atlanta; the Delta RPRC (now Tulane RPRC) in association with Tulane University at Covington, Louisiana; and the New England RPRC in association with Harvard University at Southboro, Massachusetts. In 1962, the University of California, Davis, was awarded a specialized center, designated the National Center for Primate Biology, whose function was to develop techniques for procuring, conditioning, and maintaining various nonhuman primate species. This center eventually became the California RPRC. The initial establishment of these six centers was completed in 1968, after eight years of cumulative federal funding by the NIH. These seven centers are presently supported as the RPRC program by the NCRR.

In addition to the NCRR-supported RPRC program, a number of other facilities house, breed, and maintain a variety of nonhuman primate species for biomedical research within the United States. Some of the larger facilities include: the Southwest Foundation for Biomedical Research in San Antonio, Texas; the Bowman Gray School of Medicine at Wake Forest University in Winston-Salem, North Carolina; the Laboratory for Experimental Medicine and Surgery (LEMSIP) associated with New York University in New York City; the Caribbean Primate Research Center in Puerto Rico; and the Department of Defense laboratories at a variety of locations (including the former Hollman Air Force Base, New Mexico).

Considerations for Employing Primate Species

Before the expense of human clinical trials is incurred, use of species such as nonhuman primates should be considered in the evaluation of selected candidates and formulations that may have the best chance for future studies in humans. Nonhuman primates sometimes provide important and needed animal models for human disease. The best nonhuman primate candidates are those phylogenetically closest to humans. However, cost and other considerations may preclude studies in species such as chimpanzees, orangutans, gorillas, and gibbons. These species are endangered or on the threatened species list, and they cannot be transported into the United States from their native habitats. Similar problems related to nonhuman primate research have also occurred in Europe.

The use of nonhuman primates on the endangered or threatened species list is restrictive, and it may be easier and more cost-effective to perform studies at institutions that have active breeding programs within the United States and/or facilities with capacities to perform these studies outside the United States.

The cost of studies in hominoid primates, especially chimpanzees, is prohibitive to most investigators and small companies. Institutions and facilities that perform experimentation on chimpanzees usually charge a "use or retirement fee" to assist in the long-term maintenance and care of the chimpanzee over its expected life.

Nonhuman Primates as Models for Infectious Diseases

More than 150 zoonoses—infections and diseases shared in nature by humans and other vertebrate animals—have been described and recognized.[47] Other infectious diseases can be transmitted by experimental infection of animals in a research setting.[48] Perhaps the most widely studied hominoid primate in biomedical research is the chimpanzee (*Pan* species). Chimpanzees can be experimentally infected with a number of human pathogens, and the resulting infection can serve as a model for seroconversion following infection and in some instances can induce the pathological consequences of the disease. Chimpanzees are susceptible to infection with human hepatitis A, B, and C virus (HAV, HBV, and HCV, respectively). More than 100 human cases of HAV have been associated with newly imported chimpanzees. Experimental infection with HBV results in serological and biochemical characteristics similar to those of an active infection with HBV in humans. An asymptomatic chronic HBV carrier state similar to that in humans can exist in chimpanzees and has also been reported in gorillas in the wild.[49] Chimpanzees have been invaluable in the development and testing of passive immunotherapeutic reagents and vaccines against HBV infection and subsequent disease.[50] They are an excellent predictor for human vaccine efficacy for HBV and appear to provide the only reliable animal model for studying HCV infection.

The chimpanzee has also been used in models of human immunodeficiency virus (HIV)[51] and respiratory syncytial virus (RSV) infections.[52] Chimpanzees can be infected with HIV, they seroconvert as do humans, and the virus can be isolated during infection. However, the pathological consequence of HIV infection in humans, i.e., AIDS, appears to be an extremely rare event in chimpanzees. To date, the only instance of AIDS-like symptoms reported required an incubation for more than 10 years.[53] While chimpanzees provide a good animal model to evaluate whether putative HIV vaccine candidates can induce sterile immunity and completely prevent infection, they are not adequate to determine whether a particular vaccine will prevent or delay the onset of AIDS. Housing of infected chimpanzees in specialized biosafety facilities for the long-term maintenance and care presents logistical and cost issues for investigators studying HIV infections.

Infant chimpanzees are susceptible to intranasal challenge with RSV, becoming infected and developing upper respiratory disease. However, juvenile or adult chimpanzees are not susceptible to RSV infection, presumably because of earlier exposure from animal caretakers or other chimpanzees, and they develop a subclinical infection that precludes susceptibility to an experimental infection. Chimpanzees are also susceptible to a number of bacterial and protozoal infections and have been employed in disease models for leprosy.[54,55] A number of the hominoid primates are susceptible to infection with *M. tuberculosis* (TB), and the infection of an orangutan[56] and cynomolgus monkeys with TB[57] has recently been described.

A variety of studies have examined the susceptibility of Old World and New World monkeys to infection and subsequent disease induction by human pathogens. For the most part, a variety of primate species are susceptible to infection with a number of human pathogens (see Table 2.3). The issues of pathological consequences of infection and mimicry of the human disease are not as clear-cut; hence, a number of these models are not good predictors of human disease. A number of other viral, bacterial, fungal, and protozoal agents will infect a variety of nonhuman primate species. Some of these are naturally occurring infections; others are attempts at experimental infection to develop an animal model for human disease. An excellent review of this subject matter and concerns related to biosafety and prevention of possible transmission to exposed animal handlers in contact with these infected nonhuman primates can be found elsewhere.[35] An alternative source of citations is Current Primate References, a monthly publication of the Primate Information Center, at the Regional Primate Research Center, University of Washington, Seattle. This publication provides up-to-date literature citations in specific subject areas related to nonhuman primates. A recent review of animal models to evaluate vaccines for the prevention of infectious diseases can be found elsewhere as well.[7]

Considerations in Choosing a Nonhuman Primate as an Animal Model

Comparative Immunology and Reproductive Physiology

On the basis of immunologic and reproductive physiology, the hominoid primates are most closely related to humans. Commercially available reagents that detect human immunoglobulin class and subclass, cluster of differentiation (CD) antigens, and cytokines are cross-reactive for the most part with hominoid primate analogs (Pharmingen Technical Bulletin, 1994). However, the comparative immunologic information available on nonhominoid primates is limited. Rhesus monkeys and macaque species exhibit 3 IgG subclasses,[58] baboons, chimpanzees, and humans exhibit 4 IgG subclasses.[59–61] Whether this difference is the result of relatively insensitive immunologic techniques that were employed in detecting only 3 IgG subclasses remains to be determined. More detailed com-

Table 2.3. Susceptibility of Old and New World Monkeys to Infection and Disease by Human Pathogens

Infectious Agent	Nonhuman Primate	Reference No.
Plasmodium vivax	Aotus and owl	78–80
P. falciparum	monkeys	
Hepatitis A virus	Owl monkeys	81, 82
Hepatitis A virus	African green monkeys,	83
	rhesus and cynomolgus	84
	monkeys	
Hepatitis E virus	Owl and cynomolgus monkeys	85, 86
	and tamarins	
Epstein-Barr virus	Cottontop marmosets	87, 88
Poxviruses	Variety	89, 90
Measles virus	Rhesus monkeys	91
Respiratory		
syncytial virus	Rhesus, cebus, and	52
	squirrel monkeys	
Bordetella pertussis	Common marmosets,	92, 93
	cynomolgus and rhesus	
	monkeys and Taiwan macaques	
Helicobacter pylori	Rhesus monkeys	94
Anthrax	Rhesus monkeys	95
Group B *Streptococcus*	Macaques	96, 97

parative immunologic studies have been described with baboon immunoglobulins (Igs) and murine monoclonal anti-baboon Ig reagents that cross-react with other primate Igs.[62,63] Cross-reactions between anti-human Ig reagents and other primate Igs have also been described.[64] The majority of the anti-human CD antigen reagents cross-react with Old World monkey species, although differences among species are observed.[65] Some of the human cytokine assay kits detect comparable cytokines produced in Old World primates. The closer the Old World monkey species is phylogenetically to humans (i.e., baboons versus African green monkeys), the stronger the cross-reactivity with human reagents.

For a vaccine to elicit a cellular immune response, the class I and class II products of the major histocompatibility complex (MHC) must present peptide fragments of the vaccine to T cells. Molecular characterization of primate class I and class II molecules shows that both hominoid and Old World monkey MHC molecules closely resemble their human counterparts; the expressed great ape, Old World monkey, and human MHC molecules are either direct descendants of common MHC ancestors or have evolved in a convergent manner.[66]

A major difference exists in the maternal-fetal unit among Old World monkeys. Macaque species exhibit a double discoid placentation that differs from the single discoid placentation in baboon species.[67] Human placentation is single discoid, so the baboon is more similar to the human than macaques. The uterine anatomy of macaques and baboons is also different.[67] Macaque species have a colliculus close to their cervical canal, whereas the baboons have no colliculus.

These anatomical differences preclude the routine use of the endocervical canal to gain access to the uterus in macaques. This difference may be a consideration in selecting a nonhuman primate model in studies related to maternal-fetal transfer, including maternal vaccination studies where maternal-fetal antibody transfer is of interest and *in utero* exposure to infectious agents and vertical transmission. It has also been reported in one maternal vaccination study employing a meningococcal type B glycoconjugate vaccine that pregnant rhesus monkeys and unvaccinated controls had a premature delivery rate approaching 35%, with an additional stillbirth rate approaching 20%. Although these rates appeared to be exceedingly high, they represent a logistic concern for maternal vaccination and *in utero* exposure studies. The rate of premature delivery among captive baboons in breeding colonies is less than 5%, compared to approximately 7% (less than 37 weeks gestation) in humans. Two percent of the premature baboons are low-birth-weight infants (including spontaneous abortions). An advantage of Old World monkey species, such as macaques and baboons, is the ability to time pregnancies.

Housing and Maintenance

In cost considerations for nonhuman primate studies, housing and maintenance may provide some cost relief for the budget conscious. For housing of animals, a choice may be the use of gang versus individual caging. Individual caging is more expensive to maintain, and the per diem costs per animal are higher. Social animals, such as baboons, can be housed together in gang caging, which requires less maintenance and is a cheaper alternative to individual caging. However, some animals, particularly macaques, will bite each other and fight when housed in gang cages, thus requiring additional veterinarian care during the experimental protocol. In the worst-case scenario, gang-caged animals may inflict mortal injuries on one another. Thus, while gang-cage housing may save some money, it can also jeopardize the outcome of the experiment and the experimental interpretations by the addition of variables such as stress, or possibly a decrease in the anticipated size of the experimental group. Studies that involve an infectious challenge or an infectious agent that is a human pathogen may require specialized biosafety facilities for housing and maintenance. The use of biosafety facilities in experimentation further increases the cost of housing and maintenance of nonhuman primates.

ENDOGENOUS AGENTS AS RISK FACTORS FOR ANIMAL MODELS

Because virtually all rodents used in this country for vaccine research come from commercial vendors who raise animals in barrier conditions, the threat of an investigator acquiring an infection from rodents is remote. Since many strains are also specific pathogen-free animals, exposure to endogenous agents capable of infecting humans is unlikely. Organisms that used to be potential problems for humans, such as *Streptobacillus moniliformis* (rat bite fever) LCM virus, Mycoplasma sp., Leptospira sp. and ectoparasites, are no longer common problems. It should be noted, however, that rodents infected with human pathogens are quite capable of scratching, biting and urinating on their human handlers. If the pathogen can be spread via an aerosol route, handlers may be at risk for infection secondary to animal exposure.

Nonhuman primates, like their human counterparts, harbor endogenous agents that need to be considered in the design of an experimental protocol.[68,69] While these endogenous latent viruses may not manifest as clinical disease, they present a very real hidden hazard to laboratory personnel. The risks of bites, scratches, and accidental injury is particularly a problem when working with nonhuman primates.[70,71] Even workers in the laboratory setting that utilize tissues from nonhuman primates for cell culture work are at risk.[72,73] As described above, SIV is a naturally occurring agent that infects green monkeys without causing any apparent disease. Macaque species can be infected with herpes B virus (*Herpesvirus simiae*) and become carriers of this agent. The viral agent has also been referred to as B virus, simian B or monkey B virus, cercopithicid herpesvirus, and more recently cercopithecine herpesvirus 1. This agent can cause a fatal neurologic infection in exposed animal handlers and is a risk to individuals working with materials from infected animals. The herpes B virus was first identified in a polio research scientist who died of a rapidly progressive encephalitic disease in 1932 following a bite from a macaque.[74] Herpes B virus-free macaques are available, but, to ensure their negative status, they must be obtained directly from the supplier, shipped, housed, and maintained at all times with no contact with macaques of unknown or questionable herpes B virus status. Supplies of monkeys free of herpes B virus are limited, and the cost of purchase is commensurate with their limited availability.

Baboons, although not infected with herpes B virus, can be infected with simian T-lymphotropic virus I (STLV-I),[75] a close relative of the human pathogenic virus HTLV-I. STLV-I is part of a distinct group of type C oncoviruses, which includes HTLV-I, HTVL-II, and bovine leukemia virus. Infection with STLV-I in older baboons can result in immunosuppression and lymphomas. Although these agents may be present in nonhuman primates in one colony, but not in another, investigators should be aware of their existence and determine whether or not any of these agents can affect the outcome of a protocol in a given species. The potential for cross-species transmission of these endogenous agents is a major concern for xenotransplants of nonhuman primate tissues and organs into humans.[76] For an overall review of these issues, the book *Nonhuman Primates in Biomedical Research*, edited by B. Taylor Bennett, Christian R. Abee, and Roy Henrickson and published by Academic Press is an invaluable resource.

OPTIMIZATION AND ALTERNATIVES TO THE USE OF ANIMALS

Within the last 25 years, there has been a fundamental change in how we regard the use of animals for research.[77] No longer do most investigators view animals as a sort of living test tube over which they have absolute authority. The moral and ethical responsibilities humans have with respect to the "rights" of other living species are continuously debated. An excellent discussion of this topic can be found in a book entitled *Laboratory Animals in Vaccine Production and Control,* by Henriksen[6] which details the Dutch experience, and in the more philosophic discussion by Fuchs.[3] Largely as a result of the ongoing debate, federal rules and regulations about the housing and care of animals have changed substantially, and care for all animals, including rodent species, is more humane, and greater consideration is given to how and when animals are used for experimental studies. As a natural outcome of the changes in regulations for the care of animals, the costs associated with the purchase, care, and housing of animals have increased dramatically and have become an additional restraint on the promiscuous use of animals by unenlightened investigators. Finally, the establishment of animal care and use committees at institutions performing animal research provides a mechanism for the review of research protocols in much the same way investigational review boards screen human research protocols.

The number of animals required to document a specific biologic effect should be carefully calculated in planning experiments. It is easy to determine, on the basis of expected outcome for an experimentally infected animal control versus a treatment group, the number of animals needed to achieve statistical validity. For example, if an experimental treatment, such as a vaccine, is expected to reduce disease by 50%, in an animal model where 100% of animals become infected, test and control need contain only 12 animals in order to achieve a statistical significance at $P<0.05$ according to chi square analysis. The use of techniques, such as the lethal dose in 50% of animals (LD_{50}), is no longer considered appropriate for most experiments. Such determinations can be replaced by more sophisticated statistical models that require far fewer animals. Most animal care and use committees review the number of animals requested for use and the statistical methods employed before approving the purchase of animals. In many institutions, the laboratory animal care group will provide assistance in the planning for animal experimentation.

It is also important to consider whether animals purchased for immunization studies can be used by other investigators after the immunization protocol has been completed, thereby substantially reducing procurement costs and total number of animals used. The issues regarding nonhuman primates have been discussed above, but they also include the lifetime housing costs for each primate used for research.

Alternative strategies should always be considered before a living model system is chosen. Mechanistic studies in which the immune response to a particular antigen is being assessed can often be performed *ex vivo* by obtaining the necessary cells for study from only a few animals, rather than using a large group of animals to perform *in vivo* studies. In order for this "replacement" strategy to work, surrogate markers for protection, toxicity and disease must be identified. Basic toxicologic studies with target cell lines from humans can often narrow the focus of such studies in animals. Many aspects of the immune response can also be simulated to some extent *in vitro* with use of appropriate cell cultures, particularly for viral agents where much of the basic biology can be obtained without more expensive animal experimentation. These refinements in experimental technique are exciting alternatives to the bulk use of animals for basic exploratory research.

While the reduction, replacement, and refinement strategies discussed by Henriksen[6] for vaccine development are appealing, it should also be recognized that the use of animals during preclinical and clinical trials is an important feature of the vaccine development process. It is unlikely that animals will ever be replaced completely as part of this process, but it is possible for scientists to use alternative methods and to reduce animal use by careful investigation.

REFERENCES

1. Timoney, J.F., J.H. Gillespie, F.R. Scott, and J.E. Barlough. 1988. *Hagen and Bruner's Microbiology and Infectious Diseases of Domestic Animals,* Comstock Publishing Associates, Ithaca and London.
2. Cohen, B.J. and F.M. Loew. 1984. Laboratory animal medicine historical perspectives, pp. 1–17. *In:* Cohen, B.J. and F.M. Loew (Eds.), *Laboratory Animal Medicine,* Academic Press, New York.
3. Fuchs, B.A. 1995. Use of animals in biomedical experimentation, pp. 97–127. *In:* Fuchs, B.A. (Ed.), *Scientific Integrity,* ASM Press, Washington, DC.
4. Gowen, J.W. and J. Stadler. 1967. Specificity of vaccine-conferred resistance to Salmonella typhimurium in mice. *J Infect Dis.* 117:151–161.
5. Hartman, A.B., C.J. Powell, C.L. Schultz, E.V. Oaks, and K.H. Eckels. 1991. Small-animal model to measure efficacy and immunogenicity of Shigella vaccine strains. *Infect Immun.* 59:4075–4083.
6. Hendriksen, C.F.M. 1988. *Laboratory Animals in Vaccine Production and Control,* Kluwer Academic Publishers, Dordrecht, Boston, London.
7. Makela, H., B. Mons, M. Roumiantzeff, and H. Wigzell. 1996. Animal models for vaccines to prevent infectious diseases *Vaccine.* 14:717–730.
8. Onderdonk, A.B., R.L. Cisneros, R. Finberg, J.H. Crabb, and D.L. Kasper. 1990. Animal model system for studying virulence of and host response to Bacteroides fragilis. *Rev Infect Dis.* 12(Suppl. 2):S169–177.
9. Onderdonk, A.B. 1991. Editorial response to Brook: Comparison of in vivo methods for determination of antimicrobial efficacy [editorial; comment]. *Rev Infect Dis.* 13:1181–1183.
10. Gorbach, S.L. and A. Onderdonk. 1979. Experimental animal models and human disease [editorial]. *Gastroenterology* 76:643–645.
11. Onderdonk, A.B., W.M. Weinstein, N.M. Sullivan, J.G. Bartlett, and S.L. Gorbach. 1974. Experimental intra-abdominal abscesses in rats. I. Development of animal model. *Infect. Immun.* 10:1256–1259.
12. Onderdonk, A.B., W.M. Weinstein, N.M. Sullivan, J.G. Bartlett, and S.L. Gorbach. 1974. Experimental intra-abdominal abscesses in rats. II Quantitative bacteriology of infected animals. *Infect. Immun.* 10:1256–1259.

13. Onderdonk, A.B., R.L. Cisneros, and J.G. Bartlett. 1980. Clostridium difficile in gnotobiotic mice. *Infect. Immun.* 28:277–282.

14. Onderdonk, A.B., B.R. Lowe, and J.G. Bartlett. 1979. Effect of environmental stress on Clostridium difficile toxin levels during continuous cultivation. *Appl. Environ. Microbiol.* 38:637–641.

15. Bartlett, J.G., A.B. Onderdonk, R.L. Cisneros, and D.L. Kasper. 1977. Clindamycin-associated colitis due to a toxin-producing species of Clostridium in hamsters. *J. Infect. Dis.* 136:701–705.

16. Bartlett, J.G., N. Moon, T.W. Chang, N. Taylor, and A.B. Onderdonk. 1978. Role of Clostridium difficile in antibiotic-associated pseudomembranous colitis. *Gastroenterology* 75:778–782.

17. Barnett, S.W., K.K. Murthy, B.G. Herndier, and J.A. Levy. 1994. An AIDS-like condition induced in baboons by HIV-2. *Science* 266:642–646.

18. Warren, J.T. and M.A. Levinson. 1995. Fourth annual survey of worldwide HIV, SIV, and SHIV challenge studies in nonhuman primates. *J. Med. Primatol.* 24:150–168.

19. Rodewald, A.K., A.B. Onderdonk, H.B. Warren, and D.L. Kasper. 1992. Neonatal mouse model of group B streptococcal infection. *J. Infect. Dis.* 166:635–639.

20. Finger, H., P. Emmerling, and B. Offenhammer. 1970. Reduced adjuvant activity of Bordetella pertussis vaccine in mice after priming with an immunogenic threshold dose. *Int. Arch. Allergy Appl. Immunol.* 39:45–55.

21. Folds, J., D. Walker, D. Hegarty, D. Banasiak, and J. Lang. 1983. Rocky Mountain spotted fever vaccine in an animal model. *J. Clin. Microbiol.* 18:321–326.

22. Lecornu, A. and A.N. Rowan. 1979. The use of nonhuman primates in the development and production of poliomyelitis vaccines. *Atlanta Abstr.* 7:10–19.

23. Molina, N.C. and C.D. Parker. 1990. Murine antibody response to oral infection with live aroA recombinant Salmonella dublin vaccine strains expressing filamentous hemaglutinin antigen from Bordetella pertussis. *Infect. Immun.* 58:2523–2528.

24. Poole, T. and A.W.E. Thomas. 1995. Primate Vaccine Evaluation Network (PVEN). Recommendations, Guidelines, and Information for Biomedical Research Involving Nonhuman Primates with Emphasis on Health Problems of Developing Countries.

25. Sabin, A.B. 1985. Oral poliovirus vaccine: History of its development and use and current challenge to eliminate poliovirus from the world. *J. Infect. Dis.* 151:420–436.

26. Warren, J.T. and M. Dolatshahi. 1992. Worldwide survey of AIDS vaccine challenge studies in nonhuman primates: Vaccines associated with active and passive immune protection from live virus challenge. *J. Med. Primatol.* 21:139–186.

27. Landsteiner, K. and E. Popper. 1909. Ubertragung der poliomyelitis acuta auf affen. *Z. Immunitaetsforsch. Exp. Ther.* 2:377–390.

28. Landsteiner, K. and E. Popper. 1908. Mikroscopische praparate von einem menschlichen und zwei affenrucke-marken. *Wien. Klin. Wochenschr.* 21:1830.

29. Paul, J.R. and J.D. Trask. 1932. The detection of poliomyelitis virus in so-called abortive types of the disease. *J. Exp. Med.* 45:240–253.

30. Enders, J.F., T.H. Weller, and F.C. Robbins. 1949. Cultivation of the Lansing strain of poliomyelitis virus in cultures of various human embryonic tissues. *Science* 109:85–87.

31. Salk, J.R., B.L. Bennett, L.J. Lewis, E.N. Ward, and J.S. Younger. 1953. Studies in human subjects on active immunization against poliomyelitis. 1. A preliminary report of experiments in progress. *J. Am. Med. Assoc.* 151:1081–1098.

32. Horstmann, D.M. 1985. The poliomyelitis story: A scientific hegira. *Yale J. Biol. Med.* 58:79–90.

33. U.S. Department of Health and Human Services. 1985. Guide for the Care and Use of Laboratory Animals, A Report of the Institute of Laboratory Animal Resources Committee on Care and Use of Laboratory Animals., NIH Publ. No. 86-23, Institute of Laboratory Animal Resources (ILAR).

34. U.S. Department of Health and Human Services. 1985. Guide for the Care and Use of Laboratory Animals, A report of the Institute of Laboratory Animal Resources Committee on Care and Use of Laboratory Animals., NIH Publ. No. 86-23, Institute of Laboratory Animal Resources (ILAR).

35. Adams, S.R., E. Muchmore, and J.H. Richardson. 1995. Biosafety, pp. 375–420. *In:* Adams, S.R., E. Muchmore, and J.H. Richardson (Eds.), *Nonhuman Primates in Biomedical Research, Biology and Management,* Academic Press, NY.

36. Gowen, J.W. and J.Stadler. 1967. Genetic characteristics influencing vaccine-conferred immunity. *J. Infect. Dis.* 117:129–150.

37. Persidsky, Y.E.A. 1996. Human immunodeficiency virus encephalitis in SCID mice. *Am. J. Path.* 149:1027–1053.

38. Madoff, L., L. Paoletti, J. Tai, and D. Kasper. 1994. Maternal immunization of mice with group B streptococcal type III polysaccharide-beta C protein conjugate elicits protective antibody to multiple serotypes. *J. Clin. Invest.* 94:286–292.

39. Paoletti, L.C., D.L. Kasper, F. Michon, J. DiFabio, H.J. Jennings, T.D. Tosteson, and M.R. Wessels. 1992. Effects of chain length on the immunogenicity in rabbits of group B *Streptococcus* type III oligosaccharide-tetanus toxoid conjugates. *J. Clin. Invest.* 89:203–209.

40. Fleagle J.G. 1988. Fossil Old World Monkeys, pp. 397–413. *In:* Fleagle, J.G. (Ed.), *Primate Adaptation and Evolution,* Academic Press, Inc., NY.

41. Disotell, T.R., R.L. Honeycutt, and M. Ruvolo. 1992. Mitochondiral DNA Phylogeny of the Old World Monkey Tribe Papionini. *Mol. Biol. Evol.* 9:1–13.

42. Disotell, T.R. 1994. Generic level relationships of the papionini (Cercopithecoidea). *Am. J. Phys. Anthropol.* 94:47–57.

43. Yerkes, R.M. 1932. Robert Mearns Yerkes: Psychobiologist, pp. 381–407. *In:* Yerkes, R.M. (Ed.), *A History of Psychology in Autobiography,* Clark University Press, Worcester, MA.

44. Bourne, G.H. 1971. Profile: "The Yerkes Regional Primate Research Center." *BioScience* 21:285–287.

45. Anonymous. 1968. Regional Primate Research Centers: The Creation of a Program, Publ. No. 76-1166, U.S. Department of Health, Education, and Welfare, Public Health Service, National Institutes of Health.

46. Johnson, D.K., M.L. Morin, K.A.L. Bayne, and T.L. Wolfle. 1995. Laws, Regulations, and Policies, pp. 15–31. *In:* Johnson, D.K., M.L. Morin, K.A.L. Bayne,

and T.L. Wolfle (Eds.), *Nonhuman Primates in Biomedical Research, Biology and Management*, Academic Press, NY.

47. Schultz, M.G. 1983. Emerging zoonoses. *N. Engl. J. Med.* 308:1285–1286.

48. Muchmore, E. 1987. An overview of biohazards associated with nonhuman primates. *J. Med. Primatol.* 16:55–82.

49. Linnemann, C.C., Jr., L.W. Kramer, and P.A. Askey. 1984. Familial clustering of hepatitis B infections in gorillas. *Am. J. Epidemiol.* 119:424–430.

50. Kennedy, R.C., J.W. Eichberg, R.E. Lanford, and G.R. Dreesman. 1986. Anti-idiotypic antibody vaccine·for type B viral hepatitis in chimpanzees. *Science* 232:220–223.

51. Eichberg, J.W., D.A. Lawlor, R.C. Kennedy, G.R. Dreesman, H.J. Alter, and W.C. Saxinger. 1986. Transmission of AIDS to chimpanzees: Infection, disease and immune response, pp. 443–456. *In:* Eichberg, J.W., D.A. Lawlor, R.C. Kennedy, G.R. Dreesman, H.J. Alter, and W.C. Saxinger (Eds.), *Animal Models of Retrovirus Infection and Their Relationship to AIDS*, Academic Press, NY.

52. Belshe, R.B., L.S. Richardson, W.T. London, D.S. Sly, J.H. Lorfeld, E. Camargo, D.A. Prevar, and R.M. Chanock. 1977. Experimental respiratory syncytial virus infection of four species of primates. *J. Med. Virol.* 1:157–162.

53. Anonymous. 1995. Chimp finally shows AIDS symptoms. *Science* 270:223.

54. Alford, P.L., D.R. Lee, A.A. Binhazim, G.B. Hubbard, and C.M. Matherne. 1996. Naturally acquired leprosy in two wild born chimpanzees. *Lab. Anim. Sci.* 46:611–616.

55. Hubbard, G.B., D.R. Lee, J.W. Eichberg, B.J. Gormus, K. Xu, and W.M. Meyers. 1991. Spontaneous leprosy in a chimpanzee (*Pan troglodytes*). *Vet. Pathol.* 28:546–548.

56. Shin, N.-S., S.-W. Kwon, D.-H. Han, G.-H. Bai, J. Yoon, D.S. Cheon, Y.-S. Son, K. Ahn, C. Chae, and Y-S. Lee. 1995. Mycobacterium tuberculosis infection in an orangutan (*Pongo pygmaeus*). *J. Vet. Med. Sci.* 57:951–953.

57. Walsh, G.P., E.V. Tan, E.C. Dela Cruz, R.M. Abalos, L.G. Villahermosa, L.J. Young, R.V. Cellona, J.B. Nazareno, and M.A. Horwitz. 1996. The Philippine cynomolgus monkey (*Macaca fasicularis*) provides a new nonhuman primate model of tuberculosis that resembles human disease. *Nat. Med.* 2:430–436.

58. Martin, L.N. 1982. Chromatographic fractionation of rhesus monkey (*macacca mulatta*) IgG subclasses using DEAE cellulose and protein A-sepharose. *J. Immunol. Meth.* 50:319–329.

59. Damian, R.T., N.D. Greene, and S.S. Kalter. 1971. IgG subclasses in the baboon (*Papio cynocephalus*). *J. Immunol.* 106:246–257.

60. Damian, R.T., M.F. Luker, N.D. Greene, and S.S. Kalter. 1972. The occurrence of baboon-type IgG subclass antigenic determinants within the order primates. *Folia primat.* 17:458–474.

61. Del Portillo, H.A., G.W. Schmidt, and R.T. Damian. 1987. Immunochemical analysis of baboon (*Papio cynocephalus*) IgG subclass. *Vet. Immunol. Immunopath.* 16:201–214.

62. Shearer, M.H., F.L. Stevens, H.B. Jenson, T.C. Chanh, K.D. Carey, G.L. White, and R.C. Kennedy. 1995. Crossreactions among primate immunoglobins. *Dev. Comp. Immunol.* 19:547–557.

63. Shearer, M.H., H.B. Jenson, K.D. Carey, T.C. Chanh, and R.C. Kennedy. 1994. Production and characterization of murine monoclonal antibodies specific for baboon IgG heavy and light chains. *J. Med. Primotol.* 23:382–387.

64. Black, C.M., J.S. McDougal, R.C. Holman, B.L. Evatt, and C.B. Reimer. 1993. Cross-reactivity of 75 monoclonal antibodies to human immunoglobin with sera of nonhuman primates. *Immunol. Let.* 37:207–213.

65. Rappocciolo, G., J.S. Allan, J.W. Eichberg, and T.C. Chanh. 1992. A comparative study of natural killer cell activity, lymphoproliferation and cell phenotypes in nonhuman primates. *Vet. Pathol.* 29:53–59.

66. Watkins, D.I., J. Zemmour, and P. Parham. 1993. Non-human primate MHC class I sequences. *Immunogenetics* 37:317–331.

67. Hendrickx, A.G. and W.R. Dukelow. 1995. Breeding, pp. 335–374. *In:* Hendrickx, A.G. and W.R. Dukelow (Eds.), *Nonhuman Primates in Biomedical Research, Biology and Management*, Academic Press, NY.

68. Heberling, R.L. and S.S. Kalter. 1978. Endogenous RNA oncornaviruses of nonhuman primates, pp. 87–97. *In:* Heberling, R.L. and S.S. Kalter (Eds.), *Recent Advances in Primatology*, Academic Press, London.

69. Hsiung, G.D. 1970. The major groups of simian viruses, pp. 65–81. *In:* Hsiung, G.D. (Ed.), *Infections and Immunosuppression in Subhuman Primates*, Munksgaard, Copenhagen.

70. Gerone, P.J. 1983. Biohazards and protection of personnel, pp. 187–196. *In:* Gerone, P.J. (Ed.), *Viral and Immunological Diseases in Nonhuman Primates*, Alan R. Liss, NY.

71. Muchmore, E. 1976. Health program for people in close contact with laboratory primates. *Cancer. Res. Saf. Monagr.* 2:81–99.

72. Many, B.W., C. Dykewicz, S. Fisher-Hoch, S. Ostroff, M. Tipple, and A. Sanchez. 1991. Virus zoonoses and their potential for contamination of cell cultures. *Dev. Biol. Stand.* 75:183–189.

73. Wells, D.L., S.L. Lipper, J.K. Hilliard, J.A. Steward, G.P. Holmes, and M.P. Kiley. 1989. Herpesvirus simiae contamination of primary rhesus monkey kidney cell cultures: CDC recommendations to minimize risks to laboratory personnel. *Diagn. Microbiol. Infect. Dis.* 12:333–337.

74. Sabin, A.B. and A.M. Wright. 1934. Acute ascending myelitis following a monkey bite, with the isolation of a virus capable of reproducing the disease. *J. Exp. Med.* 59:115–136.

75. Mone, J., E. Whitehead, M. Leland, G. Hubbard, and J.S. Allan. 1992. Simian T-cell leukemia virus type I infection in captive baboons. *AIDS Res. Human Retrovir.* 8:1653–1659.

76. Allan, J.S. 1996. Xenotransplantation at a crossroads: Prevention versus progress. *Nat. Med.* 2:18–21.

77. Loeb, J.M., W.R. Endee, S.J. Smith, and M.R. Schwarz. 1989. Human versus animal rights. *J. Am. Med. Assoc.* 262:2716–2720.

78. Chang, S.P., S.E. Case, W.L. Gosnell, A. Hashimoto, K.J. Kramer, L.Q. Tam, C.Q. Hashiro, C.M. Nikaido, H.L. Gipson, C.T. Lee-Ng, P.J. Barr, B.T. Yokota, and G.S.N.A. Hui. 1996. A recombinant baculovirus 42kilodalton C-terminal fragment of Plasmodium falciparum merozoite surface protein 1 protects Aotus monkeys against malaria. *Infect. Immun.* 64:253–261.

79. Geiman, Q.M. and M.J. Meagher. 1967. Susceptibility of a New World monkey to Plasmodium falciparum from man. *Nature* 215:437–439.

80. Young, M.D., J.A. Porter, and C.P. Johnson. 1966. Plasmodium vivax transmitted from monkey to man. *Science* 153:1006–1007.

81. LeDuc, J.W., S.M. Lemon, C.M. Keenan, R.R. Graham, R.H. Marchwicki, and L.N. Bionn. 1983. Experimental infection of the New World owl monkey (*Aotus trivirgatus*) with hepatitis A virus. *Infect. Immun.* 40:766–772.

82. Lemon, S.M., L.N. Binn, R. Marchwicki, P.C. Murphy, L.-H. Ping, R.W. Jansen, L.V.S. Asher, J.T. Stapleton, D.G. Taylor, and J.W. LeDuc. 1990. *In vivo* replication and reversion to wild type of a neutralization resistant antigenic variant of hepatitis A virus. *J. Infect. Dis.* 161:7–13.

83. Shevtsova, Z.V., B.A. Lapin, N.V. Doroshenko, R.I. Krilova, L.I. Korzaja, I.B. Lomovskaya, Z.N. Dzhelieva, G.K. Zairov, V.M. Stakhanova, E.G. Belova, and L.A. Sazhchenko. 1988. Spontaneous and experimental hepatitis A in Old World monkeys. *J. Med. Primatol.* 17:177–194.

84. Lankas, G.R. and R.D. Jensen. 1987. Evidence of hepatitis A virus infection in immature rhesus monkeys. *Vet. Pathol.* 24:340–344.

85. Krawczynski, K. and D.W. Bradely. 1989. Enterically transmitted non-A, non-B hepatitis: Identification of virus associated antigen in experimentally infected cynomolgus monkeys. *J. Infect. Dis.* 159:1042–1049.

86. Ticehurst, J., L.L. Rhodes, Jr., K. Krawxyznski, L.V.S. Asher, W.F. Engler, T.L. Mensing, J.D. Caudill, M.H. Sjogren, C.H. Hoke, Jr., J.W. Jeduc, J.W. Bradley, and L.N. Binn. 1992. Infection of owl monkeys (*Aotus trivirgatus*) and cynomolgus monkey (*Macaca fascicularis*) with hepatitis E virus from Mexico. *J. Infect. Dis.* 165:835–845.

87. Shope, T., D. Dechairo, and G. Miller. 1973. Malignant lymphoma in cottontop marmosets after inoculation with Epstein-Barr Virus. *Proc. Nat. Acad. Sci. USA* 70:2487–2491.

88. Young, L.S., S. Finerty, L. Brooks, F. Scullion, A.B. Rickinson, and A.J. Morgan. 1989. Epstein-Barr Virus gene expression in malignant lymphomas induced by experimental virus infection of cottontop tamarins. *J. Virol.* 63:1967–1974.

89. Espana, C. 1971. A pox disease of monkeys transmissible to man, pp. 694–708. *In:* Espana, C. (Ed.), *Medical Primatology 1970*, Karger, Basel.

90. Breman, J.G., T. Kalisa-Ruti, M.V. Steiniowski, W. Zanotto, A.I. Gromyko, and T. Arita. 1980. Human monkeypox 1970–1979. *Bull. W.H.O.* 58:165–182.

91. Hicks, J.T., J.L. Sullivan, and P. Albrecht. 1977. Immune responses during measles infection in immunosuppressed rhesus monkeys. *J. Immunol.* 119:1452–1456.

92. Huang, C., P.M. Chen, J.K. Kuo, W.H. Chiu, S.T. Lin, H.S. Lin, and Y.C. Lin. 1962. Experimental whooping cough. *N. Engl. J. Med.* 266:105–111.

93. Sauer, L.W. and L. Hambrecht. 1929. Experimental whooping cough. *Am. J. Dis. Child.* 37:732–744.

94. Fritz, D.L., N.K. Jaax, W.B. Lawrence, K.J. Davis, M.L.M. Pitt, J.W. Ezzell, and A.M. Friedlander. 1995. Pathology of experimental inhalation anthraz in the rhesus monkey. *Lab. Invest.* 73:691–702.

95. Handt, L.K., H.J. Klien, H. Rozmiarek, J.G. Fox, W.J. Pouch, and S.L. Motzel. 1995. Evaluation of two commercial serologic tests for the diagnosis of Helicobacter pylori infection in the rhesus monkey. *Lab. Animal Sci.* 45:613-617.

96. Rubens, C.E., H.V. Raff, J.C. Jackson, E.Y. Chi, J.T. Bielitzki, and S.L. Hillier. 1991. Pathophysiology and histopathology of group B streptococcal sepsis in *Macaca nemestrina* primates induced after intraamniotic inoculation: Evidence for bacterial cellular invasion. *J. Inf. Dis.* 164:320–330.

97. Larson, J.W., W.F. London, A.E. Palmer, J.W. Tossell, R.A. Bronsteen, M. Daniels, B.L. Curfman, and J.L. Sever. 1978. Experimental group B streptococcal infection in the rhesus monkey. *Am. J. Obstet. Gynecol.* 132:686–690.

3 | Validation and Standardization of Serologic Methods for Evaluation of Clinical Immune Response to Vaccines

Dace V. Madore, Nancy M. Strong, and Sally A. Quataert

PURPOSE OF ASSAY STANDARDIZATION AND VALIDATION

The ultimate test for a vaccine candidate is how it performs in its target population, humans. Since the first immunization 200 years ago by Jenner, tremendous progress has been made in the field of immunology.[1] The intricacies of the human immune response to infectious agents have been and continue to be dissected in an effort to gain an understanding of the entire process. This knowledge provides a basis for predicting the desired immune response to vaccines, which traditionally have induced responses mimicking the response to invasion by their respective pathogens. Quantitation of the immune response provides evidence to support the primary endpoint data demonstrating vaccine efficacy.[2]

Assay validation is critical in showing that a method is specific for its intended purpose and that it yields consistent and reproducible results. Standardization of the assay used to characterize the immune response to a candidate vaccine permits direct comparison of different vaccines evaluated by different laboratories. A desired outcome of using a validated assay is that a correlation between the immune response data and vaccine efficacy may be recognized and be established. Unfortunately, this clear objective is difficult to attain for most pathogens, since protection may be associated with immune responses to multiple virulence factors; in addition, qualitative differences in immune response can affect protective activity.[3-5] When a correlate between efficacy and immune response is identified, assay data can serve as a surrogate to evaluate vaccine performance at a point when it is no longer feasible or ethical to withhold immunization.[6] The data can also ensure consistent performance of a licensed vaccine and provide a means by which to evaluate a new vaccine candidate in the background of routine immunization for the same indication. The standardized test can provide a mechanism by which to assess modified vaccines (e.g., as combination products) or modified immunization schedules.[4,7,8]

What is meant by assay standardization and validation? Assay standardization is scientific acceptance of a particular procedure for generating validated results that can be used as a reference procedure to determine the consistency of results generated by other immunologic procedures. The importance of immunologic test standardization is thoroughly reviewed by Taylor et al.[9] The Food and Drug Administration's *Guidelines on the Validation of Analytical Procedures* from the International Conference on Harmonisation and Section 23 of the United States Pharmacopeia are valuable resources that define assay validation characteristics and suggest minimal testing requirements to validate each parameter.[10–12] Although the guidelines were intended to be applied to the analysis of drug products, they can also guide the validation of immunoassays and bioassays used for the analysis of clinical specimens. The distinction between immunoassays and bioassays is often not clear-cut. In this discussion, **bioassay** refers only to an assay using live cells or microorganisms, whereas **immunoassay** refers to all other methods. Characterization of essential assay parameters, including precision, accuracy, specificity, quantitation or detection limit, linearity, range, and robustness, is required to validate both immunoassays and bioassays.

Precision, the closeness of agreement between a series of measurements for the same sample in a given assay, can be considered at the levels of repeatability, intermediate precision, and reproducibility. At the first level, intraassay repeatability is determined for a sample tested multiple times on a given day by the closeness of agreement of the values obtained. In general, the within-day precision is the maximum degree of precision achievable with the assay. Introduction of the usual laboratory variables, such as different analysts, days, reagent lots, and equipment, yields the intermediate precision for a sample tested multiple times in the assay. This intermediate level of precision reflects the closeness of agreement within which results from a particular laboratory vary for the assay. While intermediate precision is determined within one laboratory, precision of the assay among laboratories, defined as reproducibility, is often determined in an interlaboratory trial. Good reproducibility among laboratories is highly desirable in a standardized method. Once a standard test method has been established, a laboratory may compare its intermediate precision with its interlaboratory reproducibility to validate assay performance.

While precision defines how well an assay can reproduce a value for a sample, it does not address the accuracy with which the assay determines that value. **Accuracy** depends on the closeness of agreement between the value determined by the assay and the value accepted as a conventional true value or an accepted reference value. When a validated and standardized assay exists for the antibodies of interest, accuracy can be determined by the statistical agreement between values obtained in a new assay and accepted values for samples in the standard method. However, for novel vaccine products, an established method and reference standard sera often do not exist and must be developed.

Critical to validation is determination of the specificity of the assay for unequivocally measuring the analyte (antibody) in the presence of all other components expected to be in the sample. With immunoassays and bioassays, specificity refers to the ability of the method to measure antibodies in clinical specimens that react with the selected antigen or microorganism. Specificity is often combined with a desired functional activity in the selection of an appropriate antigen.

During validation, additional assay parameters such as linearity, range, quantitation limit, and robustness need to be examined. For the assay method, the range of values should be established within which the measured results are directly proportional to the concentration of analyte in the sample. Combining the **linear range** with precision and accuracy, a **quantitation** limit—the lowest amount of analyte that can be quantitatively determined—can be set for the assay. Alternatively, in qualitative tests, a **detection** limit—defining the lowest amount of analyte that can be detected but not quantitated as an exact value—can be defined. For clinical immune responses quantitated or detected by immunoassays and bioassays, the quantitation or detection limit often represents the smallest amount of antibody that can be measured, given the lowest sample dilution tested and the method of reporting.

Robustness, which is the ability of the assay to supply values unaffected by small but deliberate variations in assay performance or reagents, provides assurance of the assay's reliability during normal usage. The greater the robustness, the easier it is to standardize and achieve interlaboratory reproducibility. Together, the different aspects of assay validation and standardization provide a testing method that will facilitate vaccine development.

SELECTION OF SEROLOGIC METHODS

Ideally, the appropriate immunologic methods would be identified and validated before the initiation of clinical trials. In this situation, the clinical design could accommodate the appropriate number of subjects based on the potential variation in immune response in the study population as well as the precision of the immunological method. In reality, the development and validation of immunological assays for the evaluation of a vaccine progress in concert with the development of that vaccine.

The analytical methods that will generate data predictive of vaccine performance and vaccine effectiveness must be identified. One of the challenges in the development of an assay is to have the appropriate reagents available for validation. These reagents often cannot be produced before the start of a clinical trial. For most antibody quantitation assays, the reagents include the antigens, a reference standard serum, quality control sera, and subject sera. Functional bioassays, such as bactericidal assays, opsonophagocytic assays, and virus or toxin neutralization assays, have additional requirements, such as live pathogens, complement source, fresh peripheral-blood mononuclear cells (PBMCs), or red blood cells. Irrespective of the analytic method selected, it is critical to assess precision, accuracy, specificity, linear range, quantitation limit, and robustness.

The most frequently measured parameter associated with vaccine performance is quantitation and characterization of antigen-specific serum antibodies.[13] Evidence for the prophylactic role of serum antibodies includes the protection from disease provided to young infants by maternal antibodies.[13-16] Similarly, convalescent-phase sera contain increased levels of antigen-specific antibodies that reduce both the likelihood and the severity of subsequent infections.[13,17] Passive immunoglobulin therapy with specific serum antibodies can provide protection against many bacterial and viral pathogens.[13,18-23] Particularly relevant to this chapter is the observation that active immunization results in specific serum antibody responses associated with protection.[13,24-30] Humoral neutralizing antibodies are effective because of their specificity for epitopes on the antigens of clinical importance. Another mechanism that plays a role in protection, particularly against intracellular pathogens, is cell-mediated immunity (CMI).[31-39] This chapter will not focus on the role of CMI responses to immunization but on the standardization of assays for the quantitation of serum antibodies.

The purpose and limitations of each serologic assay must be considered in light of its clinical significance.[2,4,6] Even when a quantitative immunochemical method is validated by the criteria described in the International Conference on Harmonisation guidelines, it does not demonstrate that these antibodies have biological activity for infection control or protection from disease. The results of all assays, whether immunoassays or bioassays, are surrogates for evaluation of clinical efficacy.

The analysis and interpretation of clinical data need to take into account the method used for antibody quantitation and data analysis. The method must supply specific antibody values with acceptable precision in a range that has clinical significance. It is most important to determine whether a calibrated reference standard serum that can be used for assigning accurate unitages to the results of specimen testing exists. The percentage of individuals attaining the protective level may be calculated in cases where the antibody isotype and titer associated with clinical significance have been established (e.g., preexisting titers in exposed individuals who do not succumb to disease, or protective titers derived from passive antibody administration). Where calibrated reference reagents and reliable correlates to protection are not available, a reference standard needs to be developed to provide a common denominator for all data collected. This reference standard can be assigned an arbitrary unitage or can be quantitated against precalibrated standards; methods for assigning unitage are discussed below (see Reference Standard Serum section).

The principal types of quantitative immunological methods measure either a primary manifestation of the antigen-antibody reaction, such as direct binding, or a secondary manifestation of the reaction, such as agglutination, precipitation, cell lysis, neutralization, or opsonization. The sensitivities of these tests vary greatly, and their adaptability to routine use and standardization varies.[40] No single standard immunologic method is used or recommended for evaluation of vaccines. Historically, assay methods have reflected the state of art at the time that vaccine effectiveness was demonstrated. More recently, certain methodologies, such as the enzyme-linked immunosorbent assay (EIA), have been more commonly used in vaccine evaluations for practical reasons. Compared with other serologic methods, the EIA requires smaller-volume specimens and is less hazardous (avoiding the use of live pathogens or radioisotopes), more sensitive, more convenient (easier to standardize and adapt to large numbers of specimens), and more versatile (capable of measuring isotypes as well as subclasses of antibody). Given its advantages and widespread use, this review will focus on the EIA, with emphasis on issues critical to validation and standardization. Many reviews have discussed the optimal use of this quantitative immunoassay and are recommended to the reader.[41–46]

LABORATORY PREPARATION FOR SEROLOGIC TESTING

The handling of biological human specimens, particularly sera, carries a risk of exposure of technical staff to known and unknown infectious agents. The Occupational Safety and Health Administration (OSHA) requires that hepatitis B vaccination be made available to employees handling human specimens to protect these individuals from infection with this bloodborne virus.[47] Bioassays usually involve the use of live pathogens or toxins. Thus, safety measures in the laboratory should include procedures that minimize aerosolization of clinical specimens and other potentially hazardous materials as well as appropriate biosafety-level containment. Guidelines for the proper laboratory use of pathogens are published and updated regularly by the Centers for Disease Control and Prevention (CDC).[48]

Key factors in the development and standardization of a serologic method for routine use include training of personnel, calibration and maintenance of laboratory equipment, and verification of reagent specificity and performance. Thorough documentation is essential and encompasses a validation report and a written protocol. The validation report summarizes studies validating all aspects of the assay, including assessment of reagents' specificity, determination of assay kinetics, and equipment calibration. This active document is updated as new reagents, materials (new lots), or pieces of equipment are introduced and as the assay is modified. The written protocol ensures uniformity of all steps of the assay, irrespective of personnel changes. Written protocols for proper reagent preparation, equipment calibration and maintenance, the assay itself, and data analysis procedures assure quality performance of the laboratory. In addition to written protocols, documentation should include appropriate safety guidelines for employees handling clinical specimens and encountering other biological and chemical hazards.

ASSAY DESIGN CONSIDERATIONS

While validated methods should yield consistent data over a wide range of specimens and time, intentional or accidental biases may be introduced if specimens are not handled blindly and tested randomly. The same procedural steps should be used for all specimens—reference, quality control, or subject specimens. Handling all coded specimens in the same fashion reduces the potential for technical error and ensures that control specimens reflect the precision and accuracy with which values are being assigned to subject specimens. A consistent labeling and handling procedure also facilitates the automation of the test method, which has economic benefits in larger trials.

A similarly rigorous approach must be applied to the analysis of raw data, which yields the final value associated with a specimen. Ideally, the data analysis method, whether manual or automated, should handle all specimens in the same way, without regard to the clinical protocol.

ASSAY COMPONENTS

The components in a standard EIA that determine the nature and amount of antibody binding are the antigen and the detecting secondary antibody. Several issues related to these reagents merit discussion.

Antigens

In most cases, vaccine components are the antigens used in serologic assays. It is often presumed that the antibodies being quantitated in a serologic assay are specific to a particular antigen. Unfortunately, in the development of a standardized assay, this tenet usually is not valid, and potential specific cross-reactivities must be considered. Antibodies can react with a number of antigens with different origins and chemical compositions if a common antigenic determinant group is recognized. Different antigens can share determinant groups, and such homology can be expected on the basis of the phyloantigenic relationships of microorganisms.[40] Such immunological cross-reactivity has been noted with the saccharide determinants of the polysaccharides of *Haemophilus influenzae* type b and *Escherichia coli* K-100,[49] *H. influenzae* type b and *S. pneumoniae* types 29 and 6,[50] and between *S. pneumoniae* and *Klebsiella*.[51] Moreover, some different serotypes of *S. pneumoniae* have similar sugar determinants.[52–59] Antisera to *S. pneumonia* can recognize dextrans, dextrins, hemicelluloses, and glycogens isolated from various higher plants and animals.[40] Proteins that are from different organisms but have related functions can exhibit immunological cross-reactivity.[60–63] Thus, it is desirable to use antigens that are well characterized; literature and amino acid sequence searches may be helpful in identifying potentially cross-reactive antigens. This specific cross-reactivity needs to be accommodated in the validation of any serologic assay; the degree of cross-reactivity should be quantitated and considered in the interpretation of the data.

In addition to the immunological cross-reactions intrinsic to antigenic structure and conformation, the purity and integrity of the antigen selected for the assay need to be carefully evaluated. The consistency, specificity, and accuracy of antibody quantitation can be closely related to the purity of the antigen. Because whole microorganisms express many antigens that may cross-react with antibodies in human sera, a pure antigen reflective of a vaccine's active component(s) is the most desirable. The purity and integrity of the antigen depend upon the growth and purification methods used to produce the antigen. To ensure consistency, the methods used must be shown to yield reproducible antigen preparations. Selection of the appropriate antigen—whether it be whole virus or whole bacterium, crude extracts or more purified carbohydrate, lipid or protein components—should be based on the assay requirements. Ideally, the tests are designed to quantitate antibody responses to the vaccine candidate that are relevant to disease protection.

For instance, antibodies to the serotype-specific polysaccharide capsule of *S. pneumoniae* are associated with protection from disease, but the effectiveness of antibodies to the C polysaccharide has not been established.[64,65] Purified pneumococcal polysaccharides are contaminated with various amounts of C polysaccharide and cell-wall proteins that cannot be easily removed and, when used as antigens, may bind preexisting antibodies to C polysaccharide and antibodies to proteins. Absorption of antibodies from serum with a pneumococcal absorbent preparation containing C polysaccharide but not serotype-specific polysaccharide ensures that the antipneumococcal EIA will detect only antibodies to serotype-specific pneumococcal polysaccharides.[65,66]

The relationship between the purity and the specificity of the antigen should be kept in mind throughout the designing of the test method. Inadvertent introduction of endotoxin and other impurities to the antigen may lead to overestimation of antibody levels, since human sera may contain varying levels of antibodies to the impurities. For example, Figure 3.1a shows the results of an antipneumococcal polysaccharide type 14 EIA in which one laboratory used pyrogen-free water for antigen coating and the other did not. The studies in which pyrogen-free water was not used overestimated antibody levels. Adoption of pyrogen-free water for coating by the second laboratory resulted in a significantly improved correlation of data (r=0.785 to r=0.980), as shown in Figure 3.1b. The impact of water quality on antigen coating has been observed with other antigens as well.[46]

Vaccines prepared by different methods against the same pathogen may elicit different antibody repertoires.[67] Thus, consideration must be given to the EIA antigen; the antigen that is most similar to the vaccine is likely to be the most sensitive in demonstrating a rise in titer after immunization. Increased sensitivity, however, cannot be the only criteria for acceptance. The antigen needs to represent the spectrum of potential variants of the pathogen and must be associated with functional activity (as discussed below). Thorough characterization of each antigen preparation ensures the quality of the antigen and the consistent representation of the same antigenic epitopes over time. The antibody repertoire of individuals may recognize multiple epitopes on a single antigen, and immunodominant epitopes may vary by individual. Particular epitopes may be sensitive to chemical modification (e.g., tyramination), to storage conditions, to freeze/thaw cycles, or to coating conditions.[68,69] Both enzymatic reactions and physical conditions (pH, temperature, ionic strength) as well as microbial contamination can lead to degradation of the antigen, with loss of epitope expression. An example is shown in Figure 3.2; the original level of binding to several

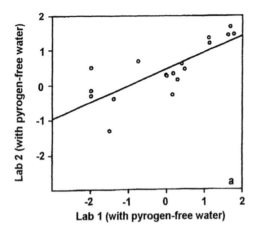
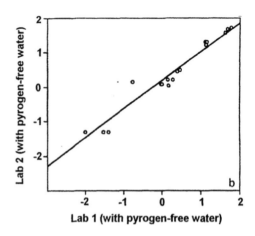

Figure 3.1. Linear regression analysis of log-transformed EIA data from Laboratory 2, which did not use pyrogen-free water for antigen coating (a), and Laboratory 1, which did use pyrogen-free water (b). Pneumococcal polysaccharide type 14 EIA was carried out according to the method of Quataert et al.,[98] with values reported as μg/mL of IgG antibody in all cases. Sera were obtained from adults before and after vaccination with a pneumococcal polysaccharide vaccine (n=19). A comparison of the regression line for (a) [0.474x + 0.466, with a correlation of r=0.785] with the regression line for (b) [0.829x + 0.203, with a correlation of r=0.980] demonstrates the improved interlaboratory correlation with the use of pyrogen-free water for coating by Laboratory 2.

epitopes of P6 protein from *H. influenzae* was reduced in a preparation stored under nonsterile conditions, whereas a sterile preparation stored at 4°C retained binding activity.

Since it is often difficult to purify a sufficient quantity of desired proteins from microorganisms, recombinantly expressed proteins are frequently considered as a source of antigens. The recombinant material needs to be compared in the assay system to the native protein expressed by the microorganism to ensure that the recombinant form presents the secondary and tertiary conformational epitopes associated with protection.

When a highly purified antigen source is desired, the purification method must be gentle enough to allow retention of the relevant antigenic structure. Often, proteins are detoxified (e.g., tetanus or diphtheria toxin). The toxoided antigen should be evaluated to ensure retention of functionally relevant epitopes; an example is shown in Figure 3.3, in which the binding of human sera to tetanus toxoid and to tetanus toxin are shown to be equivalent.

Several approaches can be taken to demonstrate that antibodies are able to bind to functional epitopes. The simplest is to test subject sera with a wide range of titers by both the EIA and a functional test (e.g., bactericidal activity, opsonization, passive protection of animals from challenge, or virus/toxin neutralization); a correlation may be evident if both assays are detecting the same populations of antibodies. An example is shown in Figure 3.4, where diphtheria antibody levels are tested by both the EIA and the toxin neutralization assay.

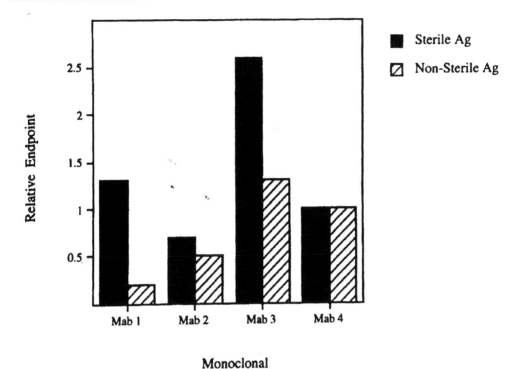

Figure 3.2. Relative reactivity in EIA of monoclonal antibodies (Mab) binding to *H. influenzae* P6 protein stored sterile or nonsterile at 4°C. Sterile antigen was bottled under good manufacturing practice conditions, while nonsterile antigen was prepared under general laboratory conditions and exhibited high-level monoclonal antibody binding before storage. The four monoclonal antibodies bind to different epitopes of the P6 molecule. In order to normalize the data and thus to allow the comparison of different antigen lots prepared at different times, endpoint dilutions at an absorbance of 0.1 are divided by the endpoint dilution observed with Mab 4, which appears to be directed to a relatively stable epitope.

Another approach is to use monoclonal antibodies that have already been characterized to bind to functional epitopes and to demonstrate that they bind specifically to the antigen used in the EIA; an example is shown in Table 3.1, where monoclonal antibodies that neutralize tetanus toxin in an *in vivo* mouse neutralization test are shown to bind to tetanus toxoid in the EIA. Antibodies in human sera blocked the binding of these tetanus toxin-neutralizing monoclonal antibodies to tetanus toxoid in the EIA. This result suggested that the same functional epitopes are present in the toxoid and the toxin (data not shown).

SERA

Human sera vary in composition, as measured by immunoglobulin isotype and subclass, and by protein and lipid content.[70-73] Infants have maternal antibodies (predominantly IgG_1) in their sera at birth, and titers decline over the first year of life.[72] While maternal antibody levels decline, infant antibodies

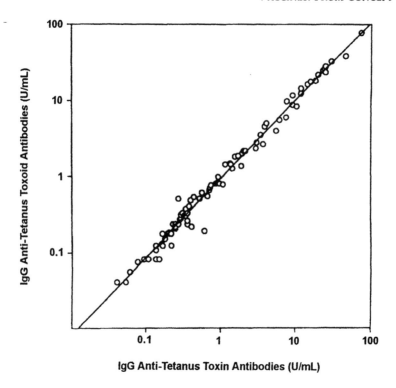

Figure 3.3. Linear regression analysis of tetanus antibody values for 91 sera tested by EIA using tetanus toxoid and tetanus toxin as antigens. These sera were from infants 2, 4, 6, and 7 months of age who had received routine childhood vaccines for diphtheria, tetanus, and pertussis (DTP); polio; and *H. influenzae* type b. The regression line for the paired endpoint data (y = 1.028x – 0.0365) and a correlation coefficient of 0.993 show an excellent linear relationship and correlation between results obtained by EIA using the two antigens.

mature at different rates, depending on isotype and subclass.[74–79] Thus, the antibody composition of pediatric serum differs from that of adult serum both qualitatively and quantitatively. The antibody repertoire of an individual matures through natural exposure and immunization to a variety of antigens and microorganisms. In addition, blood and tissues contain antibodies to surface and internal antigens of many organisms as a result of exposure through diet and environment, with no history of immunization or specific infection.[80] In certain human populations, genetics and/or high-level environmental exposure to pathogens can cause the percentages of IgM, IgG, and IgA in serum to vary from expected levels.[68,72,73,81] Healthy carriers differ from noncarriers and from vaccinees in terms of antigen-specific serum antibody levels and epitope recognition.[82,83] Immunization of individuals or repeated environmental exposure may result in the predominance of antibodies that are specific to different epitopes and that display increased affinity and avidity.[*,5,80,82,84,85] The possible heteroge-

---

* Affinity is a thermodynamic measurement of strength of non-covalent interaction between one site of the antibody and the antigen (epitope specific), while avidity charac-

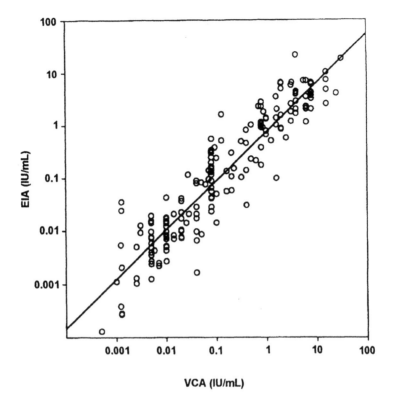

Figure 3.4. Linear regression analysis of log-transformed data from an anti-diphtheria toxoid EIA and a Vero cell diphtheria toxin neutralization bioassay (VCA) for 178 pediatric sera obtained before and after immunization. The linear regression line (y = 0.928x – 0.117) and the correlation coefficient (r=0.933) show a linear relationship and a good correlation for antibody levels measured by the two methods. The reference standard in the VCA was a World Health Organization equine serum, which was used to calibrate a human serum reference standard for use in the EIA. The value assigned to the human reference preparation was confirmed by a two-tailed paired student t-test with 178 samples analyzed in the bioassay and EIA (p=0.09).

neity of serum composition and avidity/affinity must be considered in assay validation.

In addition to inherent differences in composition, the conditions under which serum collection, processing, and retention take place can affect the quality of the specimen. Serum is often the critical component in studies whose results will determine whether a candidate vaccine will proceed into efficacy trials; thus, the integrity of serum must be preserved. Serum should be drawn and processed in a way that minimizes red blood cell lysis and the formation of precipitates.[86,87] Avoidance of routine heat treatment helps preserve specimen quality. Heat inactivation of serum can cause aggregation of proteins and nonspecific activity in an EIA.[88] Specimens to be tested in bioassays may be heat inactivated at 56°C for

terizes binding of polyclonal antisera where more than one arm of a multivalent antibody can bind to multiply expressed epitopes on an antigen.[43,84]

Table 3.1. EIA Reactivity of Monoclonal Antibodies to Tetanus Toxoid

A panel of murine monoclonal antibodies to tetanus toxin (courtesy of Dr. William Habig, U.S. FDA) reactive to both neutralizing and non-neutralizing tetanus toxin epitopes, as shown in a mouse neutralization bioassay, bind to tetanus toxoid antigen in the EIA. A level of 0.01 unit of anti-tetanus toxin/mL is neutralizing in the bioassay. Ascites fluids containing the monoclonal antibodies were tested at a dilution of 1:4000.

Monoclonal Antibody	Specificity[a,158]	Absorbance in EIA	Bioassay Neutralization Titer (Units/mL)[158]
21.83.4B	L/B	>2.00	0.030
18.2.12.6	C	>2.00	<0.001
18.1.7	C	>2.00	0.300
21.57.4	B	0.13	<0.001
21.76.10	B	>2.00	3.000
21.81.9	L	>2.00	0.030
21.18.1	L/B	0.68	<0.001
21.19.12	L/B	>2.00	<0.001
21.32.6	L/B	>2.00	0.003
21.30.3.2	B	>2.00	0.030

[a] Specificity as determined by Kenimer et al.[158]
L/B = Light Chain and Fragment B
L = Light Chain
B = Fragment B
C = Fragment C

30 minutes to reduce endogenous complement after it has been determined that such treatment will not affect the specificity or accuracy of the results.[89] Individual aliquots of serum should be handled aseptically and stored in polypropylene tubes with Teflon® ring seals to prevent lyophilization; these preparations can be stored over the long term in a scientific-grade freezer at ≤20°C. Repeated freeze/thaw cycles should be minimized so that the quality of the serum specimen is not compromised. Assessment of volume requirements for assays can help determine the appropriate aliquot size for storage.

Reference Standard Serum

The reference standard serum allows accurate assignment of unitage to the subject and quality control specimens. For many licensed vaccines and for some vaccines under development, reference standard sera are available. Sources for these sera include but are not limited to the World Health Organization, the U.S. Food and Drug Administration—Center for Biological Evaluation and Review, the U.S. Public Health Service—CDC, and various commercial organizations. When a reference standard serum is limited in supply or does not exist, it has to be prepared and "true" values assigned to it.

Selection of an appropriate reference standard serum can aid in assay validation and performance. To ensure the accuracy of values generated in an assay,
the composition of the reference standard serum and subject sera should be as
similar as possible in terms of antibody isotype and subclass distribution, affinity and avidity, specificity, and functional activity. It is difficult to prepare one
ideal reference standard serum representative of subject sera, especially when
sufficient volume cannot be readily obtained (e.g., infants). From a practical stand-
point, sera from unimmunized and immunized adults can be secured easily and
used for the preparation of a large pool that will serve as a reference standard
over a period of years. Preparation of a large volume allows the use of the reference standard to cross-standardize and validate assay methods and results in
various laboratories. Pooled sera from multiple individuals are often preferable
to sera from a single donor for reference standard preparation, providing a
broader spectrum of antibodies binding to multiple epitopes of a specific antigen and minimizing variations in isotype and subclass distribution and in antibody avidity. Generally, sera from multiple adult donors who either have
naturally acquired high levels of antibodies to the antigen of interest or have
been immunized with a vaccine related to the candidate vaccine will provide a
pool with a sufficiently high titer required for antibody quantitation in the subject population. A note of caution: Hyperimmune sera may have higher-affinity
antibodies that do not represent the subject population and that may lead to
underestimation of antibody response. Furthermore, serum specimens obtained
from any one individual at multiple time points may vary in antigen-specific
antibody level and composition. This variation necessitates continual quantitation
of serum from each collection time if the sequentially collected sera are to be
used as a reference standard.

Once a reference standard serum has been prepared, values for the antigen-
specific antibodies therein must be quantified. The values assigned to the reference standard serum serve as a basis for the consistent assignment of values to
control, subject, and replacement reference standard sera—from assay to assay
and over time. In addition, acceptance of the values assigned to the reference
standard serum facilitates its use in assay standardization across laboratories.
Assignment of an antibody value to a reference standard serum can be as easy
as the assignment of an arbitrary unitage or can be a significant undertaking (as,
for example, with weight-based assignments for specific immunoglobulin types).
While easy to assign, arbitrary unitage does not permit assessment of absolute
levels of antibody response among isotypes and subclasses or comparison of
measured responses to different antigens. Various methodologies, including
quantitative precipitation, chemical or radiological quantitation of purified antibodies specific for the antigen of interest, and equivalence of absorbance units
in parallel EIAs, have been used to assign values to reference standard sera.[28,90-97]

A quantitative precipitin approach is most easily applied to antibodies to carbohydrate antigens, since protein quantitation detects only the antibodies in the
precipitate. However, quantitation by precipitin is limited by various other factors, such as the capacity to quantitate only total immunoglobulin, the failure to

adequately precipitate lower-affinity antibodies, and the co-precipitation of immunoglobulins such as rheumatoid factors, anti-allotypes, and anti-idiotypes, which are not specific to the antigen.

Another approach to antibody quantitation is to purify the antibody of interest and then quantitate it. This method requires large amounts of serum and may necessitate the use of radioactivity; its accuracy depends upon the accuracy of the chemical or radioassay method.

Values assigned to a reference standard serum through the equivalence of absorbance units for a known amount of immunoglobulin analyzed simultaneously in parallel EIAs can be used with a variety of antigens, whether carbohydrate, protein, or lipid in nature. EIAs are capable of quantitating antibodies of specific isotypes and subclasses. The method is easily adapted for use in a homologous antigen-specific EIA if a previously calibrated human reference standard serum exists. Alternatively, a heterologous EIA, specific for antibodies to a different antigen that has a calibrated reference standard, can be used if all other assay conditions are identical. This approach was recently used to assign weight-based total Ig, IgG, IgM, and IgA values to a human antipneumococcal standard serum (lot 89-SF) for 11 pneumococcal polysaccharide serotypes.[98]

Early serum reference preparations, such as diphtheria or tetanus antitoxin, were prepared in animal species (including horses and goats) and assigned arbitrary unitage. The unitage was associated with minimum protective levels through correlation with the results of clinical efficacy trials.[13,30,99] While animal-derived reference standards are useful for bioassays such as toxin neutralization, they are inadequate for immunoassays in which the detecting reagent (secondary antibody in EIA) is specific for human sera.[100] However, human reference standard sera can be quantitated by means of a bioassay using the animal-derived reference standard. Accuracy can be confirmed by analyses of values assigned to a statistically significant number of sera tested in both the bioassay and the immunoassay. An example is shown in Figure 3.4, where a human reference standard serum for antibody to diphtheria toxoid has been quantitated using the World Health Organization's equine reference standard for diphtheria antitoxin.

Quality Control Sera

Quality control serum specimens are handled in the same manner as subject specimens (see below) and are run in every assay as a means of verifying the precision and accuracy of values assigned to subject samples. Multiple quality control sera are a critical part of assay validation and should represent the antibody isotypes and avidity of the subject sera. Ideally, several control sera are selected to represent high, median, and low antigen-specific antibody levels in the subject population. All these control sera should be treated in the same manner—with use of the same dilution scheme—as the subject sera and should be included in each assay run. Tracking of the control serum values allows establishment of the expected mean value and typical error around the mean for each control and determination of the coefficient of variation (CV) of an assay. For assays where the reported value is a noncontinuous serial dilution, such as the last positive tube in a series of tubes containing twofold sample dilutions for a bioassay, data should be log transformed prior to plotting.[9] The tracked values can be plotted as a Shewhart control chart and inspected by Western Electric rules to detect shifts or trends.[101] Control specifications established during validation are used to assess whether assay performance is within acceptable limits.[101,102]

In the validation of daily assay performance, the values for control sera provide an unbiased and objective way for a laboratory to judge whether an assay is performing as expected for a given run and whether the subjects' antigen-specific antibody values can be accepted as accurate. For EIAs, at least one control serum should be run on each assay plate along with the reference standard serum. The quantitation of the control serum, with established assay control specifications, validates the calibration of the reference curve on each plate. Differences in the antigen-binding potential of the plates and in other factors can cause within-day and between-day variation in sensitivity. In addition, since some clinical trials can span an extended period, control values can ensure the comparability of data throughout a clinical trial. Assay variation over time needs to be considered during the evaluation of the significance of differences between treatment groups. An example of long-term control data is presented in Figure 3.5, which shows the performance of a human serum control run in an EIA over a period of five years with a CV of 23%. Shifts in control data can be used to detect unexpected changes in assay performance for either identifiable or unknown reasons (e.g., reagent lot changes and unidentified changes in laboratory conditions, respectively).[101,102]

Subject Sera

Subject sera are, as their designation implies, samples from the subjects enrolled in a clinical study. Often a number of immunologic assays are needed to demonstrate the effectiveness of a candidate vaccine and to evaluate the potential interference of the candidate with the immune response elicited by existing vaccines. In some cases, the volume of blood available to meet the requirements for multiple assays will be limited—for example, when clinical trial subjects are infants. Serological methods designed to accommodate such limitations are essential. In addition, it is recommended that untreated archival aliquots of all or of representative, randomly selected clinical specimens be retained for unforeseen purposes (e.g., repeat or new tests) until the vaccine is licensed and all regulatory requirements have been met.

As discussed previously, the characteristics of reference standard sera and quality control sera should be representative of subject sera. Thus, any pretreatment of subject sera is also applied to reference standard and quality control sera to ensure consistency in assay performance. Heat treatment for complement inactivation should be limited to the aliquots that are removed for particu-

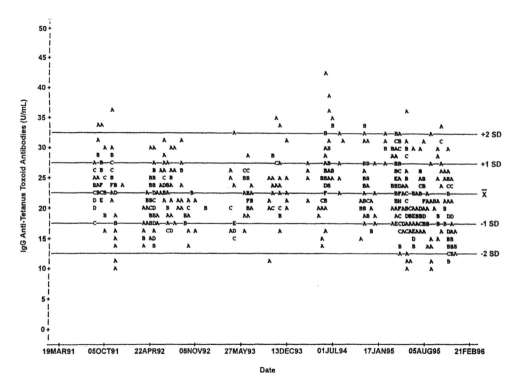

Figure 3.5. Monitoring of anti-tetanus toxoid EIA over a period of 5 years with a quality control human serum pool. Values are designated by letters denoting the number of observations (A=1, B=2, etc.). The distribution of points on either side of the mean (x) throughout the 5-year period, with the majority of the points falling within one standard deviation (SD), demonstrates the long-term assay precision (CV = 23%). The human reference standard serum used in this EIA was calibrated from human anti-tetanus immunoglobulin preparation (Miles Inc./Cutter Biological) which was previously calibrated in an *in vivo* guinea pig toxin neutralization assay employing the U.S. Standard Antitoxin and U.S. Control Antitoxin.

lar tests in which endogenous complement needs to be destroyed (e.g., toxin neutralization or other functional assays). Additional pretreatment of aliquots may be necessary to ensure specificity of antibody binding.[103] Interference of rheumatoid factor with IgG, IgA, and IgM antibody quantitation may require its removal.[104–113] Likewise, in cases where the antigen in the EIA contains contaminants, antibodies reacting with the contaminants need to be preadsorbed from the sera being tested, or the subject sera need to be subjected to further analysis (e.g., inhibition studies) to determine the proportion of the antigen-specific immune response detected by the assay.

The assay conditions need to be suitable for the quantitation of a potentially broad range of antibody titers with minimum repetition of assays, since serum volumes are limited and testing is costly. As mentioned above, the potential for assay or technician bias of results can be minimized by the testing of coded subject sera from different treatment groups in a random fashion throughout the clinical study. The precision and reproducibility of quantitative antibody assays increase when multiple dilutions are used for the determination of the antibody titer.[114] The dilutions selected should cover the range of values for subject sera and ensure consistent sensitivity at the quantitative limit. Such optimization of the assay can be difficult if the laboratory does not yet have sera from the treatment groups. Reoptimization of the dilution scheme and other assay parameters established during the development of the assay may be necessary when subject sera are received.

DETECTING ANTIBODIES

Secondary antibodies in immunoassays, including enzyme-conjugated secondary antibodies for EIA, are used to detect and quantitate antibodies in reference, quality control, and subject sera that specifically bind to the antigen. Different lots of the same reagent can vary in specificity, potency (working dilution), and shelf-life stability. The optimal dilution for use of a secondary antibody varies with each assay and needs to be determined with the antigen-specific coated plate. Antibody-binding specificity is enhanced with secondary antibodies that specifically target the immunoglobulin heavy chain. Quality control of antibody reagents detecting anti-total Ig ensures that IgG, IgM, and IgA are equally detected, as shown in Table 3.2; similarly, IgG-detecting antisera can be screened to ensure consistent detection of all of the subclasses, as shown in Table 3.3. Polyclonally-derived detecting antibodies should also be tested in the absence of subject sera to ensure that this reagent by itself does not bind to the antigen. Commercial sources of secondary antibody reagents do respect the needs of particular assays and usually cooperate in providing suitable reagents. It is preferable to purchase large quantities of an acceptable lot to minimize assay variability throughout the clinical trial and to avoid the repeated qualification testing that must be conducted with new lots.

The variable nature of polyclonal antisera has motivated investigators to use monoclonal antibodies as detecting reagents.[115,116] While this is an attractive alternative, data must be generated for each specific EIA to demonstrate that the monoclonal reagent yields the same antibody titers observed with the polyclonal reagent; supporting data should reflect a wide range of titers as well as antibodies generated under different conditions (natural infection, different vaccine formulations). The secondary monoclonal antibody reagent must detect all antibody allotypes present in the subject population and the antibody idiotypes generated in subject sera by the candidate vaccine.

ASSAY CONDITIONS

Just as assay components require characterization and qualification, the conditions for their use in a given method require evaluation. In the EIA, the optimal concentration of antigen must be identified to ensure detection of both low- and high-affinity antibodies.[41,117,118] Consistent detection of antibody responses

Table 3.2. Quality Control of Balanced Isotype Reactivity

Secondary antibody-enzyme conjugates are tested for isotype-specific binding in an EIA. Human isotype immunoglobulins are coated in separate EIA wells at 1 µg/mL. A dilution of secondary reagent that yields an absorbance of approximately 1.0 with the specific antigen under the assay conditions is tested for binding.

Enzyme Conjugate	Absorbance (405 nm) in EIA with Indicated Human Antigens		
	IgG	IgM	IgA
Goat anti-total Ig-alkaline phosphatase	1.016	0.935	0.8673
Goat anti-IgG-alkaline phosphatase	0.963	0.026	0.004
Goat anti-IgM-alkaline phosphatase	0.003	1.272	0.002
Goat anti-IgA-alkaline phosphatase	0.029	0.014	1.044

Table 3.3. Quality Control of Balanced Subclass Reactivity

Quality control of secondary antibody-enzyme for balanced reactivity with all IgG subclasses in an EIA. Purified human IgG subclass myeloma proteins are coated in separate EIA wells at 1 µg/mL. The reactivity of the anti-human enzyme conjugate is assessed in all wells at one dilution.

Enzyme-Conjugate	Absorbance (405 nm) in EIA							
	G1κ	G1λ	G2κ	G2λ	G3κ	G3λ	G4κ	G4λ
Goat anti-total Ig-alkaline phosphatase	1.62	1.48	1.14	1.55	1.28	1.73	0.88	1.62
Goat anti-IgG-alkaline phosphatase	1.26	1.23	1.24	1.21	1.33	1.64	0.99	1.30

to the antigen over time can be readily validated with a panel of sera used for this purpose whenever new reagents/materials are introduced. This panel of sera should represent the full range and all types of responses detectable by the assay. Some antigen epitopes can be altered or sterically made inaccessible during coating steps; the consistent availability of critical epitopes should be monitored either with monoclonal antibodies or with a panel of control sera.[119,120]

Precise, repeatable performance of the EIA depends on consistent antigen coating of plate wells. Different manufacturers' EIA plates can absorb reagents specifically as well as nonspecifically. If unrelated proteins (casein, bovine serum albumin, etc.) are added to the process, they may bind to the plate or to specific antibodies in some subjects' sera. The addition of detergents, such as Tween, Brij, or Triton, to buffers can eliminate the need to add irrelevant proteins to the assay buffers.[44,121]

Any immunochemical assay is a series of dynamic interactions. Incubation temperatures, humidities, and time must be kept constant throughout the assay for all sera; otherwise, the same point in the binding reaction may not be reached, and repeatability may be affected. Reproducibility is greatest when the kinetics of binding of the primary and secondary antibody, as well as substrate development, are optimized to ensure that binding has reached an equilibrium. In the EIA, plate washing between incubation steps can be critical to assay performance, affecting interwell and interplate variability. The washing procedure, whether manual or automated, must be validated to ensure equal dispensing of volumes of wash buffer, equal force of wash, and even distribution of residual buffer to all wells. For most EIA applications, if all the reagents have been qualified and the reactions optimized for equilibrium binding, the coefficient of variation of the optical density for replicate wells across a plate should not exceed 10%.

While the greatest effort in assay optimization and standardization is required during the early developmental stages, assay validation is a continual process. Changes in materials, reagents, and equipment need to be qualified both in comparison to the characteristics of the previously validated components and in relationship to overall validation of the assay. This monitoring is particularly important for assays where data are collected and collated over a long period or are compared with previously generated data.

DATA ANALYSIS

The output of an EIA is a set of optical density readings, usually of multiple dilutions of the reference standard serum, quality control sera, and subject sera. Data analysis methods compare optical density values of the standard reference serum with those of unknown specimens. The key element in the acceptability of such an analysis is a similarity of the slopes of the reference standard serum and the subject sera; parallel titration curves are critical to meaningful and reproducible assignments of values to unknown specimens. An analysis of the variation of values obtained by readings at different optical densities (i.e., precision) can help identify the optical density at which antibody value assignment varies the least. In the EIA, a lower optical density reading—closer to the quantitation limit—can minimize the effect of differences in affinity on the antibody value measured.[117,122] The method for converting optical density readings to antibody values may be programmed for computation; just like the steps of the assay method, computer programs require validation. Many programs are provided with readers or are available from public sources.[123,124] Guidance for optimization of the analysis method is described thoroughly by Plikaytis et al.[114,125] Different analysis programs yield similar values if the assays are optimized to a point where the subject, quality control, and reference standard sera behave similarly.

CORRELATIONS OF IMMUNOASSAYS TO FUNCTIONAL ASSAYS

As has already been discussed, immunoassays such as the EIA can conveniently, accurately, and reproducibly quantitate antigen-specific antibodies with small volumes of sera. However, these assays do not prove that the antibodies detected have any biological activity. While EIAs have replaced functional tests because of their practicality, it is still essential to verify that the EIA quantitates antibodies that have clinical significance. This task is best accomplished by the demonstration of a correlation between antibody values obtained by EIA and those obtained in a biologically relevant functional assay. Functional bioassays include toxin/virus neutralization assays, bactericidal assays, opsonophagocytic assays, and animal protection assays.[5,80,126–138] Because of their dependence on fresh and dynamic materials (e.g., PBMCs, live bacteria or viruses, or cell cultures), bioassays are more difficult to standardize and validate than immunoassays. Obtaining PBMCs with similar reactivity on a regular basis is an arduous and unpredictable undertaking: the cells must be used fresh, and PBMCs are a continually changing population within any one individual and between individual donors.[139,140] Efforts to develop immortal cell lines (HL-60, THP-1) may reduce this variability in some cases.[141,142]

The epitope expression or virulence of microorganisms used in bioassays can vary. Cell culture conditions and metabolic state alter antigen expression in microorganisms, affecting their immunological profile.[5,80,130,143–145] The virulence of cultured viruses and the tolerance of host cells depend on cell passage number; thus, stocks of similar passages need to be prepared to ensure consistent assay performance.[146] This sensitivity to growth conditions is consistent with the observation that laboratory strains of microorganisms may differ from wild-type strains (clinical isolates) in their expression of antigens. Thus, the antigenic target for functional tests should be carefully selected and monitored.

Similarly, many bioassays depend on unique reagents such as complement, which is highly labile and is species dependent in its activity.[5,130] It should be kept in mind that the source of complement is serum and that each serum has its own antibody repertoire and the potential for cross-reaction with the antigen in an assay.

The decision as to which bioassay is the best indicator of *in vivo* protection is fraught with pitfalls. Protective activity of antibodies in humans may not correlate with functional activity in any one *in vitro* assay, since functional assays represent a partial reconstitution of the *in vivo* immune system.[147] However, these bioassays are our best hope of capturing the functional activity of antibodies in an artificial test.

Because of the complexities and intricacies of different assays, reports of a lack of correlation between assays are not uncommon.[148–153] This lack of correlation may reflect either actual differences in the detection and quantitation of antibodies in two assays or limitations inherent in either assay. An EIA measures the antigen-specific immunoglobulins detected by the secondary antibodies (i.e., anti-total Ig, anti-IgG, anti-IgM); functional assays detect a subset of antibodies that are active under selected conditions.[5,85,120] The limitations of bioassays are numerous and usually are unique to each assay. For example, a lack of correlation of an EIA or an immunoassay to a bioassay can result if antibodies bind to epitopes that hinder antibody binding to neighboring functional epitopes. Such blocking antibodies would still have reactivity in the EIA.[80,154,155] A correlation can also be lacking if an epitope is not available or is modified in its presentation in the EIA. Antibodies may perform differently not only because of differences in epitope specificity but also because of differences in isotype, subclass, or avidity.[80,134,147,156] These variables, along with differences in assay limits and range of sensitivities, often make the clinical importance of correlations between assays difficult to interpret.

In spite of the potential difficulties in correlating data generated by two different assay methods, positive correlations can be attained if assay reagents are thoroughly evaluated and controlled.[147,156,157] An example in which the EIA serves as a surrogate for a bioassay is shown in Figure 3.4 for the quantitation of antibodies specific for diphtheria toxin.

CONCLUSIONS

It is challenging to validate and standardize a method for the evaluation of candidate vaccines because these tasks are frequently undertaken while the immune mechanism of protection in humans is still under investigation. Moreover, at this point, the antigens and epitopes critical in generating a protective immune response often have not been fully identified and characterized. Thus, assay validation and standardization proceed on the basis of past and ongoing experiences with host immune responses to infectious agents. The primary purpose of such assays is to evaluate the immune response of a target population; in contrast, the goal of clinical immunodiagnostic tests is to define the immune status of an individual. With the continual advancement in characterization of protective immune responses and the greater fine-tuning of the assays developed to evaluate these immune responses, it may become possible to use the assays to define the protective status of individuals.

Whether an assay is being developed for the evaluation of vaccines or for diagnostic purposes, it must be validated to ensure that it is specific as well as accurate with all types of specimens. Furthermore, the assay's performance must be monitored vigilantly to assure that the data it yields are consistent irrespective of operator, laboratory, or time of use. The creation and use of a reference standard serum and of quality control sera are important in meeting these objectives. Assay standardization relies on access to or development of these reagents and of the written protocols for the assay procedure.

The ultimate goal is to demonstrate the effectiveness of a candidate vaccine in disease protection or infection control in the target population; the *in vitro* assay serves as a surrogate for the demonstration of clinical effectiveness. In this capacity, *in vitro* assays may not always capture the full range of immunoresponsiveness that follows immunization. The focus of the primary assay method should be the characterization of the most common and most effective mechanism predictive of vaccine performance. As clinical trials may involve hundreds or thousands of subjects, the method needs to be practical so that responses can be evaluated in a timely and cost-effective manner.

Functional bioassays usually are not practical for large-scale clinical trials. Immunoassays like the EIA provide a more practical method for quantitation of antibodies; however, they are even further removed than bioassays from direct measurement of responses reflecting clinical performance. Several approaches to confirm the relevance of the EIA are recommended. One method is the identification of critical epitopes on the antigen used in the EIA with monoclonal antibodies that have demonstrated functional activity. Another is to evaluate representative pre- and postimmunization sera by EIA and by bioassay and to show that the kinetics of response in the target population over the immunization schedule are similar, even if the absolute unitage differs. The most common approach is to attempt to demonstrate a correlation of the response data generated by the EIA and those generated by a bioassay. However, it is often difficult to show a high degree of correlation because the two assay methods may be preferentially detecting different populations of antibodies. Antibodies that appear to be nonfunctional in a bioassay may play a protective role in forming antigen-antibody complexes targeted for clearance from the blood. Since activity in an immunoassay does serve as a surrogate for biological activity and for clinical efficacy, some correlation is necessary to justify its use.

No discussion on assay validation and standardization is complete without some mention of the importance of thorough and accurate documentation. The clinical data provided in support of a candidate vaccine's licensure are only as sound as the method used to generate the data; documentation of the assay method and its validation is essential. Accurate and reproducible vaccine-response data help lay a foundation for our further understanding and characterization of the human immune response to infection and to immunization.

REFERENCES

1. Jenner, E. 1798. An Inquiry into the Causes and Effects of the Variolae Vaccinae. London.
2. Horne, A.D. 1995. The statistical analysis of immunogenicity data in vaccine trials: A review of methodologies and issues. *Ann. N.Y. Acad. Sci.* 754:329–346.
3. Olin, P. 1995. Defining surrogate serologic tests with respect to predicting protective vaccine efficacy: Pertussis vaccination. *Ann. N.Y. Acad. Sci.* 754:273–277.
4. Decker, M.D. and K.M. Edwards. 1995. Issues in design of clinical trials of combination vaccines. *Ann. N.Y. Acad. Sci.* 754:234–240.
5. Granoff, D.M. and A.H. Lucas. 1995. Laboratory correlates of protection against *Haemophilus influenzae* type b disease: Importance of assessment of antibody avidity and immunologic memory. *Ann. N.Y. Acad. Sci.* 754:278–288.
6. Goldenthal, K.L., D.L. Burns, L.D. McVittie, B.P. Lewis, Jr., and J.C. Williams. 1995. Overview-combination vaccines and simultaneous administration: Past, present, and future. *Ann. N.Y. Acad. Sci.* 754:11–15.
7. Parkman, P.D. 1995. Combined and simultaneously administered vaccines: A brief history. *Ann. N.Y. Acad. Sci.* 754:1–7.

8. Insel, R.A. 1995. Potential alterations in immunogenicity by combining or simultaneously administering vaccine components. *Ann. N.Y. Acad. Sci.* 754:35–47.

9. Taylor, R.N., A.Y. Huong, K.M. Fulford, V.A. Przybyszewski, and T.L. Hearn. 1979. Quality control for immunologic tests. U.S. Department of Health and Human Services (CDC), Atlanta, GA.

10. U.S. Department of Health and Human Services (FDA). 1995. International conference on harmonisation; Guideline on validation of analytical procedures: Definitions and terminology availability; Notice. *Fed. Reg.* 60:11260–11262.

11. U.S. Department of Health and Human Services. 1996. International conference on harmonisation; Draft guideline on the validation of analytical procedures: Methodology; Availability. *Fed. Reg.* 61:9315–9319.

12. USPC Board of Trustees. 1995. *In:* The United States Pharmacopeia: The National Formulary, pp. 1982–1984, USP 23: NF 18. The United States Pharmacopeial Convention, Inc., Rockville, MD.

13. Robbins, J.B., R. Schneerson, and S.C. Szu. 1995. Perspective: Hypothesis: Serum IgG antibody is sufficient to confer protection against infectious diseases by inactivating the inoculum. *J. Infect. Dis.* 171:1387–1398.

14. Fothergill, L.D. and J. Wright. 1933. Influenzal meningitis: The relation of age incidence to bactericidal power of blood against causal organism. *J. Immunol.* 24:273–284.

15. Fowler, K.B., S. Stagno, R.F. Pass, W.J. Britt, T.J. Boll, and C.A. Alford. 1992. The outcome of Congenital Cytomegalovirus infection in relation to maternal antibody status. *N. Engl. J. Med.* 326:663–667.

16. Ada, G.L. 1993. Vaccines (Maternal versus neonatal immunization), 3rd-1317. *In:* W.E. Paul (Ed.) *Fundamental Immunology.* Raven Press, Ltd., NY.

17. Koff, R.S. 1993. Hepatitis B today: Clinical and diagnostic overview. *Pediatr. Infect. Dis. J.* 12:428–432.

18. Groothuis, J.R., E.A.F. Simoes, M.J. Levin, C.B. Hall, C.E. Long, W.J. Rodriguez, J. Arrobio, H.C. Meissner, D.R. Fulton, R.C. Welliver, D.A. Tristram, G.R. Siber, G.A. Prince, M. van Raden, and V.G. Hemming. 1993. Prophylactic administration of respiratory syncytial virus immune globulin to high-risk infants and young children. *N. Engl. J. Med.* 329:1524–1530.

19. Santosham, M., R. Reid, D.M. Ambrosino, M.C. Wolff, J. Almeido-Hill, C. Priehs, K.M. Aspery, S. Garrett, L. Croll, S. Foster, G. Burge, P. Page, B. Zacher, R. Moxon, B. Chir, and G.R. Siber. 1987. Prevention of *Haemophilus influenzae* type b infections in high-risk infants treated with bacterial polysaccharide immune globulin. *N. Engl. J. Med.* 317:923–929.

20. Shurin, P.A., J.M. Rehmus, C.E. Johnson, C.D. Marchant, S.A. Carlin, D.M. Super, G.F. van Hare, P.K. Jones, D.M. Ambrosino, and G.R. Siber. 1993. Bacterial polysaccharide immune globulin for prophylaxis of acute otitis media in high-risk children. *J. Pediatr.* 123:801–810.

21. Hemming, V.G., W. Rodriguez, H.W. Kim, C.D. Brandt, R.H. Parrott, B. Burch, G.A. Prince, P.A. Baron, R.J. Fink, and G. Reaman. 1987. Intravenous immunoglobulin treatment of respiratory syncytial virus infections in infants and young children. *Antimicrob. Agents Chemother.* 31:1882–1886.

22. Beasley, R.P., L.Y. Hwang, C.E. Stevens, C.C. Lin, F.J. Hsieh, K.Y. Wang, T.S. Sun, and W. Szmuness. 1983. Efficacy of hepatitis B immune globulin for prevention of perinatal transmission of the hepatitis B virus carrier state: Final report of a randomized double-blind, placebo-controlled trial. *Hepatology* 3:135–141.

23. Beasley, R.P., C.C. Lin, K.Y. Wang, F.J. Hsieh, L.Y. Hwang, C.E. Stevens, T.S. Sun, and W. Szmuness. 1981. Hepatitis B immune globulin (HBIG) efficacy in the interruption of perinatal transmission of hepatitis B virus carrier state. *Lancet* 2:388–393.

24. Peltola, H., P.H. Makela, H. Kayhty, H. Jousimies, E. Herva, K. Hallstrom, A. Sivonen, O.V. Renkonen, O. Pettay, V. Karanko, P. Ahvonen, and S. Sarna. 1977. Clinical efficacy of meningococcus group A capsular polysaccharide vaccine in children three months to five years of age. *N. Engl. J. Med.* 297:686–691.

25. Dowdle, W.R., M.T. Coleman, S.R. Mostow, and H.S. Kaye. 1973. Inactivated influenza vaccines: Laboratory indices of protection. *Postgrad. Med. J.* 49:159–163.

26. Schiffman, G. 1982. Immune response to pneumococcal vaccine: Two uses. *Clin. Immunol. News.* 3:33–37.

27. Adler, S.P., S.E. Starr, S.A. Plotkin, S.H. Hempfling, J. Buis, M.L. Manning, and A.M. Best. 1995. Immunity induced by primary human cytomegalovirus infection protects against secondary infection among women of childbearing age. *J. Infect. Dis.* 171:26–32.

28. Robbins, J.B., J.C. Parke, Jr., R. Schneerson, and J.K. Whisnant. 1973. Quantitative measurement of "natural" and immunization-induced *Haemophilus influenzae* type b capsular polysaccharide antibodies. *Pediatr. Res.* 7:103–110.

29. Madore, D.V. 1996. Impact of immunization of *Haemophilus influenzae* type b disease. *Infect. Agents. Dis.* 5:8–20.

30. U.S. Department of Health and Human Services. 1996. Recommended childhood immunization schedule-United States, January–June 1996. *MMWR* 44:940–943.

31. Zinkernagel, R.M., M.F. Bachmann, T.M. Kundig, S. Oehen, H. Pirchet, and H. Hengartner. 1996. On Immunology Memory, pp. 333–367. *In:* W.E. Paul, C.G. Fathman, and H. Metzger (Eds.). *Annual Review of Immunology*, Palo Alto, CA.

32. Schapiro, J.M., Y. Segev, L. Rannon, M. Alkan, and B. Rager-Zisman. 1990. Natural killer (NK) cell response after vaccination of volunteers with killed influenza vaccine. *J. Med. Virol.* 30:196–200.

33. Ritz, J. 1989. The role of natural killer cells in immune surveillance. *N. Engl. J. Med.* 320:1748–1749.

34. Ada, G.L. 1994. Vaccination strategies to control infections: An overview, *In: Medical Intelligence Unit: Strategies in Vaccine Design*, pp. 1–16. Ada, G.L. (Ed.), R.G. Landes Company, Austin, TX.

35. Powers, D.C. and R.B. Belshe. 1993. Effect of age on cytotoxic T lymphocyte memory as well as serum and local antibody responses elicited by inactivated influenza virus vaccine. *J. Infect. Dis.* 167:584–592.

36. Safrit, J.T. and R.A. Koup. 1995. The immunology of primary HIV infection: Which immune responses control HIV replication? *Curr. Opin. Immunol.* 7:456–461.

37. Sher, A. and R.L. Coffman. 1992. Regulation of immunity to parasites by T cells and T cell-derived cytokines, pp. 385–409. *In:* Paul, W.E., C.G. Fathman, and H. Metzger (Eds.), *Annual Review of Immunology, Vol. 10. Annual Reviews,* Palo Alto, CA.

38. Fitzgerald, T.J. 1992. Minireview: The Th1/Th2-like switch in syphilitic infection: Is it detrimental? *Infect. Immun.* 60:3475–3479.

39. Shearer, G.M. and M. Clerici. 1994. CD4+ functional T cell subsets: Their roles in infection and vaccine development, pp. 114–124. *In:* Ada, G.L. (Ed.), *Medical Intelligence Unit: Strategies in Vaccine Design.* R.G. Landes Company, Austin, TX.

40. Kwapinski, J.B.G. 1972. Antibodies, pp. 291–294. *In: Methodology of Immunochemical and Immunological Research.* John Wiley & Sons.

41. Engvall, E. and P. Pearlmann. 1972. Enzyme-linked immunosorbent assay, ELISA: III. Quantitation of specific antibodies by enzyme-labeled anti-immunoglobulin in antigen-coated tubes. *J. Immunol.* 129:129–135.

42. Engvall, E., K. Jonsson, and P. Perlmann. 1971. Enzyme-linked immunosorbent assay: Quantitative assay of protein antigen, immunoglobulin G, by means of enzyme-labelled antigen and antibody-coated tubes. *Biochim. Biophys. Acta.* 251:427–434.

43. Tijssen, P. 1985. *Laboratory Techniques in Biochemistry and Molecular Biology: Practice and Theory of Enzyme Immunoassays.* Burdon, R.H. and P.H. van Knippenberg (Eds.), Elsevier, NY.

44. Kemeny, D.M. and S.J. Challacombe. 1988. Microtitre plates and other solid phase supports, pp. 31–56. *In:* Kemeny, D.M. and S.J. Challacombe (Eds.), *ELISA and Other Solid Phase Immunoassays Theoretical and Practical Aspects,* John Wiley & Sons, NY.

45. Butler, J.E. and R.G. Hamilton. 1991. Quantitation of specific antibodies: Methods of expression, standards, solid-phase considerations, and specific applications, pp. 173–206. *In:* Butler, J.E. (Ed.), *Immunochemistry of solid-phase immunoassay,* CRC Press, Boca Raton, FL.

46. Pruslin, F.H., S.E. To, R. Winston, and T.C. Rodman. 1991. Caveats and suggestions for the ELISA. *J. Immunol. Meth.* 137:27–35.

47. U.S. Department of Labor, 29 CFR Part 1910.1030, *In:* Bloodborne Pathogens Final Rule, Occupational Safety and Health Administration, U.S. Government Printing Office, Washington, DC, 1991.

48. U.S. Department of Health and Human Services. 1989. Guidelines for prevention of transmission of human immunodeficiency virus and hepatitis B virus to health-care and public-safety workers. *MMWR* 38:1–37.

49. Insel, R.A. and P.W. Anderson, Jr. 1982. Cross-reactivity with *Escherichia coli* K100 in the human serum anticapsular antibody response to *Haemophilus influenzae* type b. *J. Immunol.* 128:1267–1270.

50. Zepp, H.D. and H.L. Hodes. 1943. Antigenic relation of type b *H. influenzae* to type 29 and type 6 pneumococci. *Proc. Soc. Exp. Biol. Med.* 52:315–317.

51. Heidelberger, M. and W. Nimmich. 1976. Immunochemical relationships between bacteria belonging to two separate families: Pneumococci and klebsiella. *Immunochemistry* 13:67–80.

52. Robbins, J.B., C.J. Lee, S.C. Rastogi, G. Schiffman, and J. Henrichsen. 1979. Comparative immunogenicity of group 6 pneumococcal type 6A(6) and type 6B(26) capsular polysaccharides. *Infect. Immun.* 26:1116–1122.

53. Heidelberger, M. and J.M. Tyler. 1964. Cross-reactions of pneumococcal types: quantitative studies with the capsular polysaccharides. *J. Exp. Med.* 120:711–719.

54. Heidelberger, M. 1962. Immunochemistry of pneumococcal types II, V, and VI; cross-reactions of type V antipneumococcal sera and their bearing on the relation between types II and V. *Arch. Biochem. Biophys.* (Suppl. 1):169–173.

55. Heidelberger, M., J.M. Davie, and R.M. Krause. 1967. Cross-reactions of the group-specific polysaccharides of streptococcal groups B and G in anti-pneumococcal sera with especial reference to type XXIII and its determinants. *J. Immunol.* 99:794–796.

56. Heidelberger, M. and P.A. Rebers. 1960. Immunochemistry of the pneumococcal Types II, V, and VI; The relation of Type VI to Type II and other correlations between chemical constitution and precipitation in antisera to Type VI. *J. Bacteriol.* 80:145–153.

57. Heidelberger, M. 1983. Precipitating cross-reactions among pneumococcal types. *Infect. Immun.* 41:1234–1244.

58. Lagergard, T. and P. Branefors. 1983. Nature of cross-reactivity between *Haemophilus influenzae* types a and b and *Streptococcus pneumoniae* types 6A and 6B. *Acta. Path. Microbiol. Immunol. Scand.* 91:371–376.

59. MacPherson, C.F.C., H.E. Alexander, and G. Leidy. 1949. Quantitative determination, in type-specific antisera to *Haemophilus influenzae*, of the antibody that cross-reacts with encapsulated pneumococci. *J. Bacteriol.* 57:443–446.

60. Zlotnick, G.W., V.T. Sanfilippo, J.A. Mattler, D.H. Kirkley, R.A. Boykins, and R.C. Seid, Jr. 1988. Purification and characterization of a peptidoglycan-associated lipoprotein from *Haemophilus influenzae*. *J. Biol. Chem.* 263:9790–9794.

61. Gotschlich, E.C., M. Seiff, and M.S. Blake. 1987. The DNA sequence of the structural gene of gonococcal protein III and the flanking region containing a repetitive sequence: homology of protein III with enterobacterial OmpA proteins. *J. Exp. Med.* 165:471–482.

62. Clements, J.D. and R.A. Finkelstein. 1979. Isolation and characterization of homogeneous heat-labile enterotoxins with high specific activity from *Escherichia coli* cultures. *Infect. Immun.* 24:760–769.

63. Dave, V.P., J.E. Allan, K.S. Slobod, F.S. Smith, K.W. Ryan, T. Takimoto, U.F. Power, A. Portner, and J.L. Hurwitz. 1994. Viral cross-reactivity and antigenic determinants recognized by human parainfluenza virus Type 1-specific cytotoxic T-cells. *Virology* 199:376–383.

64. Shapiro, E.D., A.T. Berg, R. Austrian, D. Schroeder, V. Parcells, A. Margolis, R.K. Adair, and J.D. Clemens. 1991. The protective efficacy of polyvalent pneumococcal polysaccharide vaccine. *N. Engl. J. Med.* 325:1453–1460.

65. Koskela, M. 1987. Serum antibodies to pneumococcal C polysaccharide in children: Response to acute pneumococcal otitis media or to vaccination. *Pediatr. Infect. Dis. J.* 6:519–526.

66. Siber, G.R., C. Priehs, and D.V. Madore. 1989. Standardization of antibody assays for measuring the response to pneumococcal infection and immunization. *Pediatr. Infect. Dis. J.* 8:S84–91.

67. Perrin, P., P. Versmisse, J.F. Delagneau, G. Lucas, P.E. Rollin, and P. Sureau. 1986. The influence of the type of immunosorbent on rabies antibody EIA: Advantages of purified glycoprotein over whole virus. *J. Biol. Stand.* 14:95–102.

68. Granoff, D.M., S.K. Kelsey, H.A. Bijlmer, L. van Alphen, J. Dankert, R.E. Mandress, F.H. Azmi, and R.J.P.M. Scholten. 1995. Antibody responses to the capsular polysaccharide of *Neisseria Meningitidis* Serogroup B in patients with meningococcal disease. *Clin. Diagn. Lab. Immunol.* 2:574–582.

69. Barra, A., D. Schulz, P. Aucouturier, and J.L. Preud'homme. 1988. Measurement of anti-*Haemophilus influenzae* type b capsular polysaccharide antibodies by ELISA. *J. Immunol. Methods.* 115:111–117.

70. Segal, P., P.S. Bachorik, B.M. Rifkind, and R.I. Levy. 1984. Lipids and dyslipoproteinemia, pp. 180–203. *In:* Henry, J.B. and R.A. McPherson (Eds.), *Clinical Diagnosis and Management,* 17th ed., W.B. Saunders, Philadelphia.

71. McPherson, R.A. 1984. Specific proteins, pp. 204–216. *In:* Henry, J.B. and R.A. McPherson (Eds.), *Clinical Diagnosis and Management,* 17th ed., W.B. Saunders, Philadelphia.

72. Ricardo, M.J. and R.H. Tomar. 1984. Immunoglobulins and paraproteins, pp. 860–878. *In:* Tomar, R.H., J.B. Henry, and J.V. Kadlec (Eds.), *Clinical Diagnosis and Management,* 17th ed., W.B. Saunders, Philadelphia.

73. Morell, A., F. Skvaril, and S. Barandun. 1976. Serum concentrations of IgG subclasses, pp. 37–56. *In:* Bach, F.H. and R.A. Good (Eds.), *Clinical Immunobiology,* Academic Press, NY.

74. Kanariou, M., E. Petridou, M. Liatsis, K. Revinthi, K. Mandalenaki-Lambrou, and D. Trichopoulos. 1995. Age patterns of immunoglobulins G, A & M in healthy children and the influence of breast feeding and vaccination status. *Pediatric Allergy Immunol.* 6:24–29.

75. Allansmith, M., B.H. McClellan, M. Butterworth, and J.R. Maloney. 1968. The development of immunoglobulin levels in man. *J. Pediatr.* 72:276–290.

76. Stiehm, E.R. and H.H. Fudenberg. 1966. Serum levels of immune globulins in health and disease: A survey. *Pediatrics* 37:715–727.

77. Ochs, H.D. and R.J. Wedgewood. 1987. IgG subclass deficiencies. *Ann. Rev. Med.* 38:325–340.

78. Behrman, R.E. 1992. Immunity, allergy, and diseases of inflammation, pp. 545–548. *In:* Kliegman, R.M., W.E. Nelson, and V.C. Vaughan, III (Eds.), *Nelson Textbook of Pediatrics,* 14th ed., W.B. Saunders, Philadelphia.

79. Schur, P.H., F. Rosen, and M.E. Norman. 1979. Immunoglobulin subclasses in normal children. *Pediatr. Res.* 13:181–183.

80. Taylor, P.W. 1983. Bactericidal and bacterioloytic activity of serum against gram-negative bacteria. *Microbiol. Rev.* 47:46–83.

81. Turgeon, M.L. 1996. *Immunology & Serology in Laboratory Medicine,* 2nd ed., p. 119, Mosby, NY.

82. Arakere, G. and C.E. Frasch. 1991. Specificity of antibodies to O-acetyl-positive and O-acetyl-negative group C meningococcal polysaccharides in sera from vaccinees and carriers. *Infect. Immun.* 59:4349–4356.

83. Barbour, M.L., R. Booy, D.W.M. Crook, H. Griffiths, H.M. Chapel, E.R. Moxon, and D. Mayon-White. 1993. *Haemophilus influenzae* type b carriage and immu-

nity four years after receiving the *Haemophilus influenzae* oligosaccharide-CRM197 (HbOC) conjugate vaccine. *Pediatr. Infect. Dis. J.* 12:478–484.

84. Griswold, W.R., A.H. Lucas, J.F. Bastian, and G. Garcia. 1989. Functional affinity of antibody to the *Haemophilus influenzae* type b polysaccharide. *J. Infect. Dis.* 159:1083–1087.

85. Belshe, R.B., B.S. Graham, M.C. Keefer, G.J. Gorse, P. Wright, R. Dolin, T. Matthews, K. Weinhold, D.P. Bolognesi, R. Sposto, D.M. Stablein, T. Twaddell, P.W. Berman, T. Gregory, A.E. Izu, M.C. Walker, and P. Fast. 1994. Neutralizing antibodies to HIV-1 in seronegative volunteers immunized with recombinant gp120 From the MN strain of HIV-1. *JAMA* 272:475–480.

86. *Manual of Clinical Laboratory Immunology*, 4th ed. p. 429. *In:* Rose, N.R., E.C. de Macario, J.L. Fahey, H. Friedman, and G.M. Penn (Eds.), ASM, Washington DC, 1992.

87. Linke, E.G. and J.B. Henry. 1984. Clinical pathology/laboratory medicine purposes and practice. *In:* Henry, J.B. and R.A. McPherson (Eds.), *Clinical Diagnosis and Management*, 17th ed., W.B. Saunders, Philadelphia.

88. Tada, Y., T. Ishikawa, M. Chazono, I. Yoshida, and R. Nii. 1991. Effect of serum treatment on pertussis antibody determination by ELISA. *Dev. Biol. Stand.* 73:175–184.

89. Turgeon, M.L. 1996. *Immunology and Serology in Laboratory Medicine*, 2nd ed., pp. 118–119, Mosby, NY.

90. Kabat, E.A. and M.M. Mayer. 1961. *Kabat and Mayer's Experimental Immunochemistry*, 2nd ed., pp. 848–849, Charles C. Thomas, Springfield, IL.

91. Gotschlich, E.C., M. Rey, R. Triau, and K.J. Sparks. 1972. Quantitative determination of the human immune response to immunization with meningococcal vaccines. *J. Clin. Invest.* 51:89–96.

92. Rudolph, K.M. and A.J. Parkinson. 1994. Measurement of pneumococcal capsular polysaccharide serotype-specific immunoglobulin G in human serum, a method for assigning weight-based units to proposed reference sera. *Clin. Diagn. Lab. Immunol.* 1:526–530.

93. Seppala, I.J.T., N. Routonen, A. Sarnesto, P.A. Mattila, and O. Makela. 1984. The percentages of six immunoglobulin isotypes in human antibodies to tetanus toxoid: Standardization of isotype-specific second antibodies in solid-phase assay. *Eur. J. Immunol.* 14:868–875.

94. Shackelford, P.G., D.M. Granoff, S.J. Nelson, M.G. Scott, D.S. Smith, and M.H. Nahm. 1987. Subclass distribution of human antibodies to *Haemophilus influenzae* type b capsular polysaccharide. *J. Immunol.* 138:587–592.

95. Zollinger, W.D. and J.W. Boslego. 1981. A general approach to standardization of the solid-phase radioimmunoassay for quantitation of class-specific antibodies. *J. Immunol. Methods.* 46:129–140.

96. Makela, O. and F. Peterfy. 1983. Standard sera in solid-phase immunoassays. *Eur. J. Immunol.* 13:815–819.

97. Fomsgaard, A. and B. Dinesen. 1987. ELISA for human IgG and IgM antilipopolysaccharide antibodies with indirect standardization. *J. Immunoassay.* 8:333–350.

98. Quataert, S.A., C.S. Kirch, L.J.Q. Wiedl, D.C. Phipps, S. Strohmeyer, C.O. Cimino, J. Skuse, and D.V. Madore. 1995. Assignment of weight-based anti-

body units to a human antipneumococcal standard reference serum, Lot 89-S. *Clin. Diagn. Lab. Immunol.* 2:590–597.

99. U.S. Department of Health and Human Services. 1991. Diphtheria, tetanus, and pertussis: Recommendations for vaccine use and other preventive measures. *MMWR* 40:1–28.

100. Heron, I. 1994. Workshop II: Clinical evaluation and surveillance of combined vaccines (definition of minimum standards for immunogenicity-potential problems of serological correlates-possible interference; Serological correlates for diphtheria-tetanus-whole cell pertussis vaccines). *Biologicals* 22:389–390.

101. Western Electric Co., Inc. 1956. *Statistical Quality Control Handbook*, 11th ed.

102. Montgomery, D.C. 1985. *Introduction to Statistical Quality Control*, 2nd ed. John Wiley & Sons, NY.

103. Feldman, H.A. 1968. Removal by heparin-MnC12 of nonspecific rubella hemagglutinin serum inhibitor (32743). *P.S.E.B.M.* 127:570–573.

104. Muller, F., M. Moskophidis, and H.L. Borkhardt. 1987. Detection of immunoglobulin M antibodies to *Treponema pallidum* in a modified enzyme-linked immunosorbent assay. *Eur. J. Clin. Microbiol.* 6:35–39.

105. Yolken, R.H. and P.J. Stopa. 1979. Analysis of nonspecific reactions in enzyme-linked immunosorbent assay testing for human rotavirus. *J. Clin. Microbiol.* 10:703–707.

106. Cerny, E.H., C.E. Farshy, E.F. Hunter, and S.A. Larsen. 1985. Rheumatoid factor in syphilis. *J. Clin. Microbiol.* 22:89–94.

107. Salonen, E.M., A. Vaheri, J. Suni, and O. Wager. 1980. Rheumatoid factor in acute viral infections: Interference with determination of IgM, IgG, and IgA antibodies in an enzyme immunoassay. *J. Infect. Dis.* 142:250–255.

108. Angarano, G., L. Monno, T.A. Santantonio, and G. Pastore. 1984. New principle for the simultaneous detection of total and immunoglobulin M antibodies applied to the measurement of antibody to hepatitis B core antigen. *J. Clin. Microbiol.* 19:905–910.

109. Krishna, R.V., O.H. Meurman, T. Ziegler, and U.H. Krech. 1980. Solid-Phase enzyme immunoassay for determination of antibodies to cytomegalovirus. *J. Clin. Microbiol.* 12:46–51.

110. Joassin, L. and M. Reginster. 1986. Elimination of nonspecific cytomegalovirus immunoglobulin M activities in the enzyme-linked immunosorbent assay by using anti-human immunoglobulin G. *J. Clin. Microbiol.* 23:576–581.

111. Hamilton, R.G. and N.F. Adkinson, Jr. 1988. Quantitative aspects of solid phase immunoassays, pp. 57–84. *In:* Kemeny, D.M. and S.J. Challacombe (Eds.), *ELISA and Other Solid Phase Immunoassays: Theoretical and Practical Aspects*, John Wiley & Sons, NY.

112. Martins, T.B., T.D. Jaskowski, C.L. Mouritsen, and H.R. Hill. 1995. An evaluation of the effectiveness of three immunoglobulin G (IgG) removal procedures for routine IgM serological testing. *Clin. Diagn. Lab. Immunol.* 2:98–103.

113. Verkooyen, R.P., M.A. Hazenberg, G.H. van Haaren, J.M. van den Bosch, R.J. Snijder, H.P. van Helden, and H.A. Verbrugh. 1992. Age-related interference with *Chlamydia pneumoniae* microimmunofluorescence serology due to circulating rheumatoid factor. *J. Clin. Microbiol.* 30:1287–1290.

114. Plikaytis, B.D., P.F. Holder, L.B. Pais, S.E. Maslanka, L.L. Gheesling, and G.M. Carlone. 1994. Determination of parallelism and nonparallelism in bioassay dilution curves. *J. Clin. Microbiol.* 32:2441–2447.

115. Jefferis, R., C.B. Reimer, F. Skvaril, G. de Lange, N.R. Ling, J. Lowe, M.R. Walker, D.J. Phillips, C.H. Aloisio, T.W. Wells, J.P. Vaerman, C.G. Magnusson, H. Kubagawa, M. Cooper, F. Vartdal, B. Vandvik, J.J. Haaijman, O. Makela, A. Sarnesto, Z. Lando, J. Gergely, E. Rajnavolgyi, G. Laszlo, J. Radl, and G.A. Molinaro. 1985. Evaluation of monoclonal antibodies having specificity for human IgG sub-classes: Results of an IUIS/WHO collaborative study. *Immunol. Lett.* 10:223–252.

116. Jefferis, R., C.B. Reimer, F. Skvaril, G.G. de Lange, D.M. Goodall, T.L. Bentley, D.J. Phillips, A. Vlug, S. Harada, J. Radl, E. Claassen, J.A. Boersma, and J. Coolen. 1992. Evaluation of monoclonal antibodies having specificity for human IgG subclasses: Results of the 2nd IUIS/WHO collaborative study. *Immunol. Lett.* 31:143–168.

117. Griswold, W.R. 1987. Theoretical analysis of the sensitivity of the solid phase antibody assay (ELISA). *Mol. Immunol.* 24:1291–1294.

118. Devey, M.E. and M.W. Steward. 1988. The role of antibody affinity in the performance of solid phase assays, pp. 135–153. *In:* Kemeny, D.M. and S.J. Challacombe (Eds.) *ELISA and Other Solid Phase Immunoassays: Theoretical and Practical Aspects,* John Wiley, NY.

119. Butler, J.E. 1991. Perspectives, configurations and principles, pp. 3–26. *In:* Butler, J.E. (Ed.), *Immunochemistry of Solid-Phase Immunoassay,* CRC Press, Boca Raton, FL.

120. Sanchez-Pescador, L., P. Paz, D. Navarro, L. Pereira, and S. Kohl. 1992. Epitopes of herpes simplex virus type 1 glycoprotein B that bind type-common neutralizing antibodies elicit type-specific antibody-dependent cellular cytotoxicity. *J. Infect. Dis.* 166:623–627.

121. Brown, W.R., S.E. Dierks, J.E. Butler, and J.M. Gershoni. 1991. Immunoblotting: membrane filters as the solid phase for immunoassays, pp. 151–172. *In:* Butler, J.E. (Ed.), *Immunochemistry of Solid-Phase Immunoassay,* CRC Press, Boca Raton, FL.

122. Hebert, C.N., S. Edwards, S. Bushnell, P.C. Jones, and C.T. Perry. 1985. Establishment of a statistical base for use of ELISA in diagnostic serology for infectious bovine rhinotracheitis. *J. Biol. Stand.* 13:245–253.

123. Kirkwood, T.B.L., V.A. Seagroatt, and S.J. Smith. 1986. Statistical aspects of the planning and analysis of collaborative studies on biological standards. *J. Biol. Stand.* 14:273–287.

124. Peterman, J.H. 1991. A summary of the principal immunoassay data analysis packages. *In: Immunochemistry of Solid-Phase Immunoassay,* pp. 293–297. Butler, J.E. (Ed.), CRC Press, Boca Raton, FL.

125. Plikaytis, B.D., S.H. Turner, L.L. Gheesling, and G.M. Carlone. 1991. Comparisons of standard curve-fitting methods to quantitate *Neisseria meningitidis* group A polysaccharide antibody levels by enzyme-linked immunosorbent assay. *J. Clin. Microbiol.* 29:1439–1446.

126. Gillenius, P., E. Jaatmaa, P. Askelof, M. Granstrom, and M. Tiru. 1985. The standardization of an assay for pertussis toxin and antitoxin in microplate culture of Chinese hamster ovary cells. *J. Biol. Stand.* 13:61–66.

127. Forsgren, M. 1985. Standardization of techniques and reagents for the study of rubella antibody. *Rev. Infect. Dis.* 7:S129–132.

128. Wood, D.J. and A.B. Heath. 1992. The second international standard for anti-poliovirus sera types 1, 2 and 3. *Biologicals* 20:203–211.

129. Relyveld, E., N.H. Oato, M. Huet, and R.K. Gupta. 1992. Determination of antibodies to pertussis toxin in working reference preparations of anti-pertussis sera from various national control laboratories. *Biologicals* 20:67–71.

130. Mandrell, R.E., F.H. Azmi, and D.M. Granoff. 1995. Complement-mediated bactericidal activity of human antibodies to poly $\alpha2\rightarrow8$ n-acetylneuraminic acid, the capsular polysaccharide of *Neisseria meningitidis* serogroup B. *J. Infect. Dis.* 172:1279–1289.

131. Schneerson, R., L.P. Rodrigues, J.C. Parke, Jr., and J.B. Robbins. 1971. Immunity to disease caused by *Hemophilus influenzae* type b II. Specificity and some biologic characteristics of "natural," infection-acquired, and immunization-induced antibodies to the capsular polysaccharide of *hemophilus influenzae* type b. *J. Immunol.* 107:1081–1089.

132. Smith, D.H., G. Peter, D.L. Ingram, A.L. Harding, and P. Anderson. 1973. Responses of children immunized with the capsular polysaccharide of *Hemophilus influenzae* type b. *Pediatrics* 52:637–644.

133. Gray, B.M. 1990. Opsonophagocidal activity in sera from infants and children immunized with *Hemophilus influenzae* type b conjugate vaccine (meningococcal protein conjugate). *Pediatrics* 85:694–697.

134. Vioarsson, G., I. Jonsdottir, S. Jonsson, and H. Valdimarsson. 1994. Opsonization and antibodies to capsular and cell wall polysaccharides of *Streptococcus pneumoniae*. *J. Infect. Dis.* 170:592–599.

135. Atanasiu, P. 1973. Quantitative assay and potency test of antirabies serum and immunoglobulin. *In: Laboratory Techniques in Rabies*, 3rd ed., Vol. 23, pp. 314–318. World Health Organization, Geneva.

136. Fine, D.P., J.L. Kirk, G. Schiffman, J.E. Schweinle, and J.C. Guckian. 1988. Analysis of humoral and phagocytic defenses against *Streptococcus pneumoniae* serotypes 1 and 3. *J. Lab. Clin. Med.* 112:487–497.

137. Shurin, P.A., G.S. Giebink, D.L. Wegman, D. Ambrosino, J. Rholl, M. Overman, T. Bauer, and G.R. Siber. 1988. Prevention of pneumococcal otitis media in chinchillas with human bacterial polysaccharide immune globulin. *J. Clin. Microbiol.* 26:755–759.

138. Smith, A.L., D.H. Smith, D.R. Averill, Jr., J. Marino, and E.R. Moxon. 1973. Production of *Haemophilus influenzae* b meningitis in infant rats by intraperitoneal inoculation. *Infect. Immun.* 8:278–290.

139. El-Daher, N., J.E. Nichols, and N.J. Roberts, Jr. 1994. Analysis of human antiviral cytotoxic T-lymphocyte responses for vaccine trials using cryopreserved mononuclear leukocytes: Demonstration of feasibility with influenza virus-specific responses. *Clin. Diagn. Lab. Immunol.* 1:487–492.

140. Roberts, N.J., Jr. 1980. Variability of results of lymphocyte transformation assays in normal human volunteers. *Am. J. Clin. Pathol.* 73:160–164.

141. Breitman, T.R. 1990. Growth and differentiation of human myeloid leukemia cell line HL60. *Methods. Enzymol.* 190:118–131.

142. Tsuchiya, S., M. Yamabe, Y. Yamaguchi, Y. Kobayashi, T. Konno, and K. Tada. 1980. Establishment and characterization of a human acute monocytic leukemia cell line (THP-1). *Int. J. Cancer* 26:171–176.

143. Anderson, P., J. Pitt, and D.H. Smith. 1976. Synthesis and release of polyribophosphate by *Haemophilus influenzae* type b *in vitro*. *Infect. Immun.* 13:581–589.

144. Dolin, R. 1995. Human studies in the development of human immunodeficiency virus vaccines. *J. Infect. Dis.* 172:1175–1183.

145. Kuratana, M., M.R. Loeb, E.J. Hansen, and P. Anderson. 1989. The antigenic specificity of a serum factor-induced phenotypic shift in *Haemophilus influenzae* type b, strain Eag. *J. Infect. Dis.* 159:1135–1138.

146. Wiedbrauk, D.L. and S.L.G. Johnston. 1993. General laboratory procedures: Quality assurance, pp. 45–53. In: *Manual of Clinical Virology*, Raven Press, NY.

147. Meade, B.D., F. Lynn, G.F. Reed, C.M. Mink, T.A. Romani, A. Deforest, and M.A. Deloria. 1995. Relationships between functional assays and enzyme immunoassays as measurements of responses to acellular and whole-cell pertussis vaccines. *Pediatrics* 96:595–600.

148. Balachandran, N., S. Bacchetti, and W.E. Rawls. 1982. Protection against lethal challenge of BALB/c mice by passive transfer of monoclonal antibodies to five glycoproteins of herpes simplex virus type 2. *Infect. Immun.* 37:1132–1137.

149. Mills, E.L., F.F. Arhin, F. Moreau, B. Tapiero, J.W. Coulton, and D.L. Moore. 1995. Antibody levels by immunoglobulin class in children aged 6–60 months immunized with *Neisseria meningitidis* group C polysaccharide, Abstract #408, p. 118. In: Abstracts of the 33rd Annual Meeting for Infectious Diseases Society of America. Infectious Diseases Society of America, Washington, D.C.

150. Milagres, L.G., S.R. Ramos, C.T. Sacchi, C.E.A. Melles, V.S.D. Vieira, H. Sato, G.S. Brito, J.C. Moraes, and C.E. Frasch. 1994. Immune response of Brazilian children to a *Neisseria meningitidis* serogroup B outer membrane protein vaccine: Comparison with efficacy. *Infect. Immun.* 62:4419–4424.

151. Siber, G.R., J. Leszczynski, V. Pena-Cruz, C. Ferren-Gardner, R. Anderson, V.G. Hemming, E.E. Walsh, J. Burns, K. McIntosh, R. Gonin, and L.J. Anderson. 1992. Protective activity of a human respiratory syncytial virus immune globulin prepared from donors screened by microneutralization assay. *J. Infect. Dis.* 165:456–463.

152. Devash, Y., J.R. Rusche, and P.L. Nara. 1990–1991. Methods for analysis of biologically functional antibodies to the HIV-1 gp120 principal neutralizing domain. *Biotechnol. Ther.* 2:49–62.

153. Ward, R.L., A.Z. Kapikian, K.M. Goldberg, D.R. Knowlton, M.W. Watson, and R. Rappaport. 1996. Serum rotavirus neutralizing-antibody titers compared by plaque reduction and enzyme-linked immunosorbent assay-based neutralization assays. *J. Clin. Microbiol.* 34:983–985.

154. McCutchan, J.A., D. Katzenstein, D. Norquist, G. Chikami, A. Wunderlich, and A.I. Braude. 1978. Role of blocking antibody in disseminated gonococcal infection. *J. Immunol.* 121:1884–1888.

155. Griffiss, J.M. 1975. Bactericidal activity of meningococcal antisera blocking by IgA of lytic antibody in human convalescent sera. *J. Immunol.* 114:1779–1784.
156. Aase, A., G. Bjune, E.A. Hoiby, E. Rosenqvist, A.K. Pedersen, and T.E. Michaelsen. 1995. Comparison among opsonic activity, antimeningococcal immunoglobulin G response, and serum bactericidal activity against meningococci in sera from vaccinees after immunization with a serogroup B outer membrane vesicle vaccine. *Infect. Immun.* 63:3531–3536.
157. Belshe, R.B., L.P. van Voris, M.A. Mufson, E.B. Buynak, A.A. McLean, and M.A. Hilleman. 1982. Comparison of enzyme-linked immunosorbent assay and neutralization techniques for measurement of antibody to respiratory syncytial virus: Implications for parenteral immunization with live virus vaccine. *Infect. Immun.* 37:160–165.
158. Kenimer, J., W. Habig, and C. Hardegree. 1983. Monoclonal antibodies as probes of tetanus toxin structure and function. *Infect. Immun.* 42:942–948.

4

Considerations in the Production of Vaccines for Use in Phase 1 Clinical Trials and Preparation of the Manufacturer's Protocol

Lawrence C. Paoletti

INTRODUCTION

This chapter outlines many aspects of the production of a Phase 1 lot of vaccine and of the assembly of the manufacturer's part of an IND submission. It is designed to familiarize the researcher with standard operating procedures, considerations in the generation of the vaccine, potency and stability testing, and storage of the final bottled vaccine. This chapter has been written with the understanding that, although not all vaccines follow the same path of production, there are issues such as scale-up, formulation, and bottling that are common to all vaccines. The information included herein is targeted for use by researchers working in non-GMP (good manufacturing practices) conditions. GMP facilities follow more rigorously designed standard operating procedures (SOPs),[1] adhere to strict validation programs for reagents and equipment, and employ quality control and regulatory personnel to assist in preparation of documents.

The decision to produce the initial clinical lot of vaccine in the laboratory where the research and development work was accomplished is both wise and practical. However, this endeavor represents a transition of the basic researcher to a production manager, with an accompanying mind-shift from that of problem solver to that of careful technician. To complete this task successfully, the laboratory is transformed into a production site where the usual experimentation is temporarily halted while the vaccine is generated. Restrictions on the use of shared equipment in the laboratory should be implemented to reduce the risk of contamination. It also may be necessary to obtain new purification equipment and reagents, and hire personnel to complete the task in a timely fashion.

Much of the preparation for initial vaccine production can be reviewed in a pre-IND meeting with the FDA, which will cover not only the roles and expectations of the FDA (see Chapter 6), but also what should be included in a completed IND application. It is strongly recommended that the researcher arrange a pre-IND meeting before beginning to produce a clinical lot of vaccine in the research laboratory.

STANDARD OPERATING PROCEDURES (SOPs)

SOPs are method protocols used in a research laboratory. Unlike generic protocols, they contain specific details and information usually missing from routine laboratory notebooks: source and lot number of reagents, specific step-by-step instructions on how to perform the test, and space for the date and signature of the person(s) who performed the task. Although no standard format exists for an SOP, some of its essential components are outlined in Table 4.1. Some laboratories have even made the effort to design an SOP for writing SOPs—a precaution that ensures uniformity as well as continuity in format as laboratory personnel changes.

A standardized SOP should contain:

- title of procedure;
- description of methods sufficiently detailed that they can easily be repeated by another scientist;
- manufacturer and lot number(s) of reagents and disposable equipment (e.g., filters) used;
- signature or initials of the person(s) who performed the procedure;
- methods of analysis, including mathematical and statistical calculations;
- deposition of the original and final report.

The original or "master" SOP should be maintained by the principal investigator to ensure that he or she is aware of and has approved all changes that are made to the SOP. Preparation, updating and compliance of SOPs should be the responsibility of the principal investigator. SOPs are essential building blocks for the production of a vaccine, are included in the IND application, and are critical to the FDA as documentation of the generation of the vaccine.[1] Most academic research facilities can generate vaccines only under good laboratory practices (GLPs), but the SOPs developed at these sites can ultimately be adapted and expanded by a larger group or an industrial partner that could manufacture the vaccine within the guidelines of the FDA's GMPs. Therefore, neither time nor effort should be spared in generating and maintaining SOPs.

GENERATING A LOT OF VACCINE FOR A
PHASE 1 CLINICAL TRIAL

With the completion of SOPs for every aspect of vaccine production and post-production testing, the laboratory is ready to manufacture a lot of vaccine that will be evaluated for safety and immunogenicity in a Phase 1 clinical trial. Because Phase 1 trials are performed with 30 or fewer subjects, it would seem that a small batch of vaccine should suffice. However, approximately half of the final product will be consumed in bulk analysis, final container filling, and post-bottling testing. This level of depletion should be considered in the calculations of the scale-up production of vaccine.

Table 4.1. Essential Components of a Standard Operating Procedure

I. Title of method
   - Date assay is performed and signature of person(s) conducting experiment
   - Complete citation of original publication
II. Materials
   - Reagents, including manufacturer, grade, and lot numbers
   - Equipment, including size (e.g., Erlenmeyer flask - 250 mL) and quantity needed
   - Special conditions (e.g., chemical hood, CO_2 incubator), specific incubation periods as necessary
III. Protocol
   - Steps numbered in logical fashion, with substeps as necessary
   - Data handling, analysis, and filing of original data
   - Results, including any problems encountered with the assay
IV. Safety considerations
   - List of all safety concerns, special precautions, institutional and OSHA regulations related to the method and description of the location in the laboratory of the material safety data sheets (MSDSs) that accompany the chemicals used in the method
V. Revisions
   - Date of revision and initials of both revisor and the principal investigator, with updated copies of the method in a binder containing all of the SOPs

Vaccine production follows a logical series of events beginning with the purchase of equipment if scale-up is necessary. All equipment should be thoroughly cleaned, sterilized, and reserved exclusively for use in vaccine production. If growth media contain animal products, certification of the country of origin and validation of preparation are needed. This information can be obtained from the manufacturer and will be included in the IND application. If the procedure begins with the culturing of a microorganism, it is prudent to limit the work performed in that laboratory solely to that organism. Similarly, use of shared equipment and laboratory traffic should be reduced to a minimum or avoided in order to reduce the risk of contamination during the manufacture of the vaccine.

Reagents such as chromatography matrices, buffers, and media should be prepared carefully and sterilized. Routine checks for sterility should be performed on all reagents and the results recorded in the SOP. If reagents need to be further purified or regenerated (e.g., as with ion exchange resins), results of the regeneration and the procedure itself should also be recorded.

Production procedures should follow the SOP as closely as possible. If changes are made while a task is being performed, a notation should be made in the margins explaining the reason for the change, and initialed by the person making the change. A second laboratory worker should confirm and initial the change.

BULK VACCINE

After the final steps in preparation have been completed and before the vaccine is bottled, the material is referred to as the bulk vaccine. A lot number assigned to the vaccine can reflect the stage of production. For example, lot #1-97B denotes that this is the first vaccine prepared in 1997 and that it is in the bulk stage of production. The bulk vaccine should undergo a battery of tests before it is released for filling, but it is important to note that these tests do not substitute for general safety and sterility analyses performed later on the final bottled product. Analyses of the bulk vaccines should include tests for sterility, purity, identity, unwanted substances in the final product, true weight or equivalent measure of the material, and potency.

Sterility

Microbial sterility of the bulk vaccine should be performed by incubation of a sample on a solid or liquid medium for 24 to 48 hours at 37°C. Results of this test should be documented.

Purity

Purity of the bulk material should be evaluated by methods best suited to the particular vaccine. For example, if it is a polysaccharide-protein conjugate, separate measurements for total carbohydrate and total protein may be a reasonable way to evaluate purity. If necessary, these measurements can be augmented by specific inhibition reactions with mono-specific antiserum and pure standards.

Identity

Qualitative assessments of the active component of the vaccine will serve as a test for identity. Specific competition reactions described above may also serve to identify antigenically the active component of the vaccine.

Contaminants

The complete removal of unwanted reagents used during vaccine production should be evaluated in the bulk material. For example, if ammonium sulfate was used in generating the vaccine, documentation of the removal of ammonia and of sulfates will be necessary. It is counterproductive to allow the vaccine to be bottled if, at this stage, it contains unwanted materials.

Quantity

A measure of the amount of material present is imperative for calculating the dilution of the bulk material required for the final container fill. For some vaccines this measure is expressed in mass units usually determined on lyophilized material.

Potency

Evaluation of potency is a requirement for all vaccines. Potency is usually assessed with bulk vaccine preparations. Tests for potency consist of either *in vivo* or *in vitro* measures of the product to elicit a given result, which should, in turn, be supportive of the efficacy of the vaccine in humans. For example, a vaccine against typhoid (Typhoid Vaccine Live Oral Ty21a, Swiss Serum and Vaccine Institute, Berne, Switzerland) is delivered in a capsule containing cells of attenuated *Salmonella typhi*. Each capsule should contain between 2×10^9 and 6×10^9 CFU of viable attenuated *S. typhi*, a range that has correlated well with protective responses in humans.[2]

In vivo tests of potency have been refined in an effort to reduce the number of animals required (see Chapter 2). *In vitro* tests have been developed and standardized and the correlation with *in vivo* tests established for evaluating tetanus and diphtheria toxoid vaccines.[3,4] Precise physicochemical analysis (i.e., molecular size, degree of cross-linking, protein-to-carbohydrate ratio) of vaccines may also replace animals as a measure of potency, a technique used with *Hemophilus influenzae* type b polysaccharide-protein conjugate vaccines.[5]

FINAL CONTAINER FILL

Several decisions need to be made before the vaccine is bottled: formulation, number of doses per vial, inclusion of an adjuvant, label information, storage of vaccine, and evaluation of potency (Table 4.2). When contemplating bottling issues, it may be helpful to review the formulation of previously developed and licensed vaccines. For some vaccines, this information can be found in the *Physicians' Desk Reference*.

Formulation

Initial lots of vaccine should be formulated on the basis of the researcher's own experience. That is, if the investigator can document that an experimental vaccine maintained as a liquid preparation remained immunogenic or efficacious in animals over a period of time when stored at a particular temperature, then this formulation may be a practical starting point. However, to some degree, the formulation depends on the method of storage and the stability of the vaccine in an aqueous or lyophilized condition. Lyophilized vaccines may contain a filler such as sucrose or lactose to add bulk to the preparation. These fillers are usually added to the mixture at a working range of 2 to 5% by weight. Procedures to lyophilize vaccines, determine residual moisture content of the dried vaccine, and discussions of important issues regarding this procedure are de-

Table 4.2. Details of Final Vaccine Bottling

I.   Form of vaccine
   - Liquid
     - What diluent will be used (e.g., 0.9% saline or PBS)?
     - Will an adjuvant be incorporated in the preparation?
   - Dry
     - Is a stabilizer/filler (e.g., sucrose) required?
     - The moisture content should be measured.
     - How will the vaccine be reconstituted? Will a special diluent need to be prepared?
     - Can a commercially available diluent be used with your vaccine?
II. Number of doses per vial
   - Single dose
     - Is a bacteriostatic preservative (e.g., thimerosal) required?
   - Multiple doses
     - Is preservative required?
     - How many doses per vial?
III. Adjuvant
   - If an adjuvant is included, will it be bottled separately or admixed in the formulation?
   - What is the minimum amount of adjuvant needed?
IV. Labeling information
V.   Storage
   - Where and under what conditions will the final product be stored?
VI. Plan for periodic evaluation of potency and stability.

tailed by Adams.[6] It is imperative that the solubility of the lyophilized vaccine with the appropriate diluent be determined. If a commercially available diluent will not be used, then one will need to be prepared under GMP conditions. As with the vaccine, the diluent used for reconstitution must pass sterility and general safety tests.

Container Size/Doses

Initial clinical lots of vaccine are most often manufactured on a small scale, and thus the decision of whether the vaccine should be packaged as a single or as a multidose preparation may be based on practicality. Although there is less loss in filling multidose vials (both liquid or lyophilized forms), a bacteriostatic agent may need to be incorporated to a liquid formulation or to the diluent used to reconstitute lyophilized vaccines. Also, the moisture contents of lyophilized vaccines should be measured. If a liquid formulation is used, single-dose bottling may be a more prudent initial approach. In any case, the firm contracted to bottle the vaccine must have the proper bottling equipment to meet the manufacturer's needs.

Adjuvants

The use of an adjuvant may or may not play a part in an initial clinical evaluation; however, it is never too early to consider how it may be incorporated into the vaccine formulation. Adjuvants may be adsorbed to the vaccine during production or admixed immediately before delivery. Although several new adjuvants are under development, the only ones allowed for use with licensed vaccines in the United States are aluminum hydroxide and aluminum phosphate.[7] Vaccines can be adsorbed with aluminum during the purification process or admixed with an aluminum-containing diluent during reconstitution. Physical parameters important in adsorption of antigens to aluminum compounds and methods used to determine optimal adsorption have been published.[8,9]

Labeling

Each vial of vaccine should be clearly labeled with the following information: vaccine, diluent, amount of the active component per volume dose, storage instructions, caution notification (i.e., CAUTION: NEW DRUG LIMITED BY FEDERAL LAW TO INVESTIGATIONAL USE), site of manufacture, auspices under which the vaccine was prepared, and lot designation, control, and vial number (Figure 4.1).

Storage Conditions and Temperature

Vaccine should be stored at a temperature that will preserve its stability and potency. Most vaccines are stable for years at 0 to 8°C, but others (i.e., some viral preparations) require lower temperatures. A record of the storage conditions and temperature should be kept on file by the retainer of the bottled vaccine.

Stability

Examination of the stability of the vaccine maintained under different conditions in the appropriate diluent is a part of product development. Criteria that constitute measures of stability should be proposed by the researcher, especially if the antigen has unique properties. For example, it is known that the presence of sialic acid is critical in the maintenance of an antigenically important epitope in vaccines constructed from the capsular polysaccharides of group B *Streptococcus*.[10,11] Thus, monitoring of the degree of sialylation of the polysaccharide over time may constitute a test of stability.

It is also important to assess the temperature at which the vaccine can be maintained. Stability of a vaccine can be tested by employing "accelerated stability" temperature regimens.[12] For these studies, bottled vaccines are maintained at 4°C, 37°C, and a higher temperature such as 45°C for 1, 2, 6, and 12 months

GBS type III polysaccharide-TT vaccine in 0.05 M PBS containing 0.01% thimerosal

54 μg polysaccharide/0.5 ml. Store at 2 - 8°C.

CAUTION: NEW DRUG LIMITED BY FEDERAL LAW TO INVESTIGATIONAL USE.

University Med. Sch. - DMID.NIAID.NIH

Lot: #95-4s   900045-012

Figure 4.1. Example of a vaccine label.

and evaluated for stability at each time point. It may be convenient to perform tests for stability and potency simultaneously.

PREPARATION OF THE MANUFACTURER'S PROTOCOL OF AN IND APPLICATION

An IND application is a multicomponent document prepared by the research team that contains all of the vaccine manufacturing protocols and plans for clinical evaluation. This document must be submitted and approved by the FDA before clinical trials can begin.

The manufacturer's protocol, also referred to as the chemistry, manufacturing, and control data section, constitutes the bulk of the IND application. Key components of a manufacturer's protocol are listed in Table 4.3.

Vaccine Preparation

The vaccine preparation section contains the step-by-step methods used to prepare the particular lot of vaccine. It should follow a sequential order of events in production and be partitioned to highlight natural breaks or stopping points in production. For example, if the vaccine is a glycoprotein conjugate, description of the purification of the carbohydrate should be separated from that of the purification of the protein. The manufacturing protocol should begin with a description of the strain(s) used to prepare the component(s) of the vaccine. If the strain was obtained from an outside source, indicate how the strain was verified for authenticity and purity. Maintenance of the seed strain(s), preparation of the growth media, fermentation conditions used, and harvesting of the organism and/or culture fluids should be detailed.

SOPs that describe the purification of the antigen, from crude material to the final purified form, are included, as are in-line processing data such as wet weight of cells, amount of crude antigen recovered, and temporary storage conditions. Graphs, tables, scans, tracings, photographs generated during production should be included and clearly referred to in the text; step-by-step mathematical equations and statistical analysis used to arrive at values; and, finally, the results of

Table 4.3. Contents of the Manufacturer's Protocol

  I. Vaccine preparation
   - Growth of the organism
   - Purification of antigen
   - Chemical and biochemical analysis of purified antigen
 II. Summary of composition of vaccine
III. References
 IV. Appendixes
   - Standard operating procedures used in preparing the vaccine
   - Supporting documentation such as certification of products, compositional analysis, scans

chemical and immunological assays performed on the final purified vaccine are also included.

Summary

The next segment of the manufacturer's protocol is a summary of the characteristics (e.g., pH, molecular size, purity, endotoxin level) of the vaccine. This overview should include the composition of the bulk and final (bottled) vaccine products.

References

This section contains complete citations to published work cited in the text and complete copies of unpublished or in-press findings.

Appendices

Routine SOPs (i.e., protein or nucleic acid assays) and supporting material such as a chemical manufacturer's certificates of purity should make up the appendix. Certificates of purity can be obtained directly from the manufacturer of the product. Be sure that the appropriate lot, batch, or control number is requested.

SUMMARY AND SUBMISSION OF IND APPLICATION

Although most of the vaccine preparation is performed by laboratory researchers, outside contractors are usually required to complete specific tasks under GMP conditions. Outside contractors with expertise in specific areas such as endotoxin and heavy-metal testing, bottling, and testing of sterility and general safety are essential in vaccine production. A summary of typical events in the production of a vaccine (manufactured at an academic institution) is outlined (Figure 4.2). Documentation of findings from outside contractors should be in-

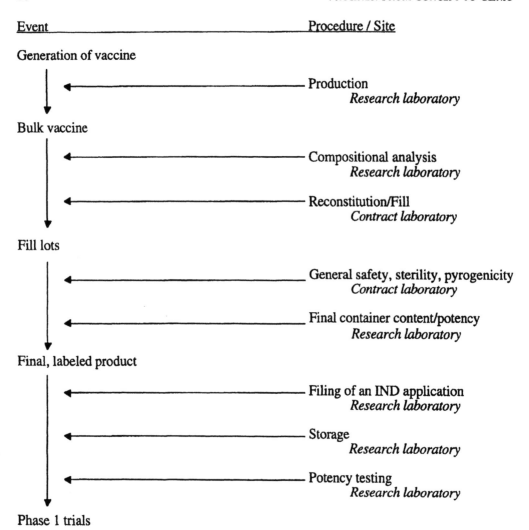

Figure 4.2. Flow diagram of events for the preparation of vaccine for Phase 1 clinical testing.

cluded in the manufacturer's protocol section of the IND application. The completed manufacturer's protocol, along with a clinical protocol (see Chapter 5) is incorporated into the IND application and submitted to the FDA for approval. The FDA has 30 days from the date of receipt to review the application. A clinical hold can be imposed and the trial cannot begin if questions are posed to the manufacturer. Satisfactory response to the clinical-hold queries will result in an approved IND submission.

ACKNOWLEDGMENTS

I would like to thank Drs. Arthur O. Tzianabos and Gerald B. Pier for critical review of this manuscript.

REFERENCES

1. DeSain, C. 1993. Documentation basics that support good manufacturing practices, Advanstar Communications, Cleveland.
2. Habig, W.H. 1993. Potency testing of bacterial vaccines for human use. *Vet. Microbiol.* 37:343–351.
3. Gupta, R.K., S. Higham, C.K. Gupta, B. Rost, and G. Siber. 1994. Suitability of the Vero cell method for titration of diphtheria antitoxin in the United States potency test for diphtheria toxoid toxoid. *Biologicals* 22:65–72.
4. Manghi, M.A., M.F. Pasetti, M. Brero, S. Deluchi, G. di Paola, V. Mathet, and P.V. Eriksson. 1994. Development of an ELISA for measuring the activity of tetanus toxoid in vaccines and comparison with the toxin neutralization test in mice. *J. Immunol. Methods* 168:17–24.
5. Gupta, R.K. and G.R. Siber. 1995. Reappraisal of existing methods for potency testing of vaccines against tetanus and diphtheria. *Vaccine* 13:965–966.
6. Adams, G. and D, J. 1996. Lyophilization of vaccines, pp. 167–185. *In:* Robinson, A., G.H. Farrar, and C.N. Wiblin (Eds.), *Vaccine Protocols.* Humana Press, Inc., Totowa, NJ.
7. Gupta, R.K. and G.R. Siber. 1995. Adjuvants for human vaccines-current status, problems and future prospects. *Vaccine* 13:1263–1276.
8. Hem, S.L. and J.L. White. 1995. Structure and properties of aluminum-containing adjuvants, pp. 249–276. *In:* Hem, S.L. and J.L. White (Eds.), *Vaccine Design: The Subunit and Adjuvant Approach.* Plenum Press, NY.
9. Lindblad, E.B. 1995. Aluminium adjuvants. *In:* Lindblad, E.B. (Ed.), *The Theory and Practical Application of Adjuvants.* John Wiley & Sons Ltd., NY.
10. Wessels, M.R., C.E. Rubens, V.J. Benedi, and D.L. Kasper. 1989. Definition of a bacterial virulence factor: Sialylation of the group B streptococcal capsule. *Proc. Natl. Acad. Sci. USA* 86:8983–8987.
11. Kasper, D.L., C.J. Baker, R.S. Baltimore, J.H. Crabb, G. Schiffman, and H.J. Jennings. 1979. Immunodeterminant specificity of human immunity to type III group B *Streptococcus. J. Exp. Med.* 149:327–339.
12. Galazka, A. 1989. Loss of potency of vaccines at elevated temperatures. *Vaccine* 7:479.

5 | Helpful Hints for Preparing an IND Application

Martha J. Mattheis and Pamela M. McInnes

INTRODUCTION

As soon as a candidate vaccine antigen is identified, the investigator/manufacturer should begin planning the series of steps that need to be taken en route to clinical evaluation. Documentation of each of these steps will form the basis of the Investigational New Drug (IND) Application to the Center for Biologics Evaluation and Research (CBER) of the Food and Drug Administration (FDA). An IND application is a request to the FDA to be allowed to administer an investigational new drug to humans. The FDA does not "approve" INDs. Instead it acknowledges receipt of the application and, if it does not disapprove, (that is, place the application on "clinical hold" because of safety issues), the study may proceed. This "lack of disapproval" must be secured prior to interstate shipment and administration of any investigational vaccine. Drugs include vaccines, and throughout this chapter the two terms may be used interchangeably.

Many investigators incorrectly assume that a vaccine has to be manufactured in a licensed facility before its clinical testing can begin. Although that is probably the more common scenario, a vaccine prepared in an academic research setting can be used in early clinical trials provided guidelines are followed in the preparation of "clinical-quality" vaccine and appropriate preclinical testing is performed. Prerequisites for moving a vaccine from the laboratory to clinical trial include purity and preclinical safety and immunogenicity. These qualities must be demonstrated, documented, and validated, where appropriate, before the outset of clinical testing.

CLINICAL TRIAL DEFINITIONS

In designing the clinical evaluation path for any investigational vaccine, it is important to delineate the target population for the vaccine. For example, if the target population is infants, then the clinical path will begin with testing in healthy adults and will continue with older children and then with infants. Depending on the vaccine, dosing issues, including the number of doses and the concentrations to be administered for maximal immunogenicity and safety, can frequently be defined only in the target population.

Phase 1 Trial

The first clinical trial of an investigational vaccine is termed the Phase 1 trial. The purpose of a Phase 1 trial is to evaluate the clinical safety of the vaccine. This trial is usually conducted in a small number of healthy adults. The number of subjects enrolled may vary, but usually ranges from 10 to 40. For parenterally administered vaccines, detailed information is gathered on the local and systemic responses to the vaccine, including pain, redness, swelling, and fever. The evaluations extend over several days after immunization. Information is generally collected on prototype case-report forms or recorded by subjects in symptom diaries. Along with safety data, initial immunogenicity data may be gathered in the Phase 1 trial, and these data will guide the dose selections for the next clinical trial. With some vaccines, special studies (e.g., measurement of the shedding of live vaccine organisms) should be implemented during early Phase 1 testing.

Phase 2 Trials

The purpose of the Phase 2 trial is to evaluate the effect of dose-ranging on the safety and immunogenicity of the vaccine. The number of subjects in a Phase 2 trial will be larger than the number in a Phase 1 trial but usually not sufficiently large for an assessment of efficacy. Sometimes the lines between Phase 1 and Phase 2 trials are blurred, and the two are blended into what is designated a Phase 1/2 trial. This term generally describes a small clinical trial evaluating safety, immunogenicity, and (in many instances) dose effects of the vaccine.

Phase 3 and Phase 4 Trials

Also called "efficacy trials," Phase 3 trials are designed to provide the pivotal efficacy data required for licensure of a vaccine. Phase 4 trials involve large-scale postmarketing surveillance. Their purpose is to obtain data on adverse events occurring with very low frequency. Neither a Phase 3 nor a Phase 4 trial would be undertaken with vaccine produced in an academic setting. These trials will not be described further in this chapter.

PROCEDURES AND REQUIREMENTS FOR FILING AN IND APPLICATION

There are standard procedures and requirements for filing IND applications and for the use of investigational new vaccines. Investigators must become informed consumers if they are to navigate the complex, but manageable, regulatory issues associated with an IND application. The relevant regulatory requirements can be found in the Code of Federal Regulations (CFR), Title 21, Part 312. Investigators planning to file an IND application need to be familiar with the intent and language of these regulations and of other regulations found in Title 21 (Table 5.1).

SOURCES OF INFORMATION

Information about procedures and requirements can be obtained from several sources. The most practical method is to call CBER, FDA (301/827-3070) and request an IND information packet. This packet contains forms needed for the application, relevant regulations, a list of available guidelines and points to consider in the development of human biological products (see below), and helpful reprints. Answers to general questions about IND submissions can also be obtained from the above CBER telephone number or from an alternative telephone number (301/827-2000). In addition, information is available via the FDA's home page (http://www.fda.gov) and via direct automated fax lines to CBER (301/827-3844 and 301/827-3156) and to the Center for Drugs Evaluation and Research (CDER) (301/827-0577). Lists of relevant documents are available from the automated fax lines and specific documents on those lists may be requested. The CFR itself can be purchased by writing to the Superintendent of Documents, Government Printing Office, Washington DC 20402, or by calling the Government Printing Office (202/783-3238). The investigator/manufacturer can be aided by documents titled Guidelines, Guidance, International Conference on Harmonization (ICH), Federal Register and Points to Consider (PTC) which are available from both CBER and CDER. Guidelines articulate procedures or standards of general applicability that, although not legal requirements, are acceptable to the FDA. Many of the guidelines issued by CDER for drugs are also relevant to biologics, including vaccines. Some of these guidelines were prepared many years ago and more recent documents, not necessarily referred to as guidelines, are available on some of the same subjects. For example, the guideline entitled "Submitting Documentation for the Stability of Human Drugs and Biologics," was prepared in February 1987. The FDA published a final guideline entitled "Quality of Biotechnological Products: Stability Testing of Biotechnological/Biological Products" in the Federal Register for July 10, 1996. This guideline was prepared by the ICH. Earlier versions on stability testing are available from the FDA either as ICH documents or as Federal Register publications.

Among the documents titled "Guidance" is "Guidance for Industry - Content and Format of Investigational New Drug Applications (INDs) for Phase 1 Studies of Drugs, Including Well-Characterized, Therapeutic, Biotechnology-Derived Products." Points to Consider documents suggest issues that must be considered with regard to a given topic. These documents are kept current by periodic updating. For example, the document on the production and testing of new drugs and biologicals produced by recombinant DNA technology was issued in No-

Table 5.1. Regulations (from Title 21, Code of Federal Regulations) Relevant to an IND Application

Part 25	Environmental Impact Considerations
Part 50	Protection of Human Subjects
Part 56	Institutional Review Boards (IRBs)
Part 58	Good Laboratory Practice for Nonclinical Laboratory Studies (GLP)
Parts 210 and 211	Current Good Manufacturing Practice (GMP) in Manufacturing, Processing, Packing, or Holding of Drugs
Part 312	Investigational New Drug Application
Part 600	Biological Products
Part 610	General Biological Products Standards
Subpart 314.126	Adequate and Well-Controlled Clinical Studies

vember 1983, revised and updated in April 1985, and supplemented in April 1992 by the addition of information on nucleic acid characterization and genetic stability.

PRE-IND MEETING

In addition to written information, a pre-IND meeting with the FDA may be requested. For investigator/manufacturers, such a meeting is advisable, because it provides a forum for a discussion of and questions about the proposed IND filing with knowledgeable FDA members. To maximize the usefulness of the meeting, the investigator must provide, at a minimum, a draft clinical protocol, manufacturing plans and specific written questions.

PREPARING THE IND APPLICATION

In the package of IND information available from the FDA are two key forms: form FDA 1571, which provides the structure for the IND submission, and form FDA 1572, which deals with investigator-related issues. The initial IND application, and each subsequent submission should be accompanied by a completed form FDA 1571, whose front and back pages are reproduced in Figures 5.1 and 5.2. Some suggestions for completing form FDA 1571 and for compiling the IND application follow, with a focus on parts of the application that can be confusing or troublesome.

Box 1: Only one individual or organization should be designated as the sponsor. An IND may be held or sponsored by a pharmaceutical company, a private, academic, or other organization, or an individual. A sponsor/investigator is an individual who both initiates and conducts a clinical investigation and under whose immediate direction the investigational drug is being administered or dispensed.

Boxes 2-5: Self-explanatory.

DEPARTMENT OF HEALTH AND HUMAN SERVICES PUBLIC HEALTH SERVICE FOOD AND DRUG ADMINISTRATION **INVESTIGATIONAL NEW DRUG APPLICATION (IND)** *(TITLE 21, CODE OF FEDERAL REGULATIONS (CFR) Part 312)*	*Form Approved: OMB No. 0910-0014.* *Expiration Date: November 30, 1995.* *See OMB Statement on Reverse*
	NOTE: No drug may be shipped or clinical investigation begun until an IND for that investigation is in effect (21 CFR 312 40)

1. NAME OF SPONSOR Dr. John Doe	2. DATE OF SUBMISSION XX-XX-XXXX

3. ADDRESS *(Number, Street, City, State and Zip Code)* Brigham & Women's Hospital The Channing Laboratory 181 Longwood Avenue Boston, MA 02115	4. TELEPHONE NUMBER *(Include Area Code)* (617) xxx-xxxx

5. NAME(S) OF DRUG *(Include all available names: Trade, Generic, Chemical, Code)* Type III Group B streptococcal polysaccharide-tetanus toxoid conjugate vaccine	6. IND NUMBER *(if previously assigned)*

7. INDICATION(S) *(Covered by this submission)*
Prevention of Group B streptococcal disease

8. PHASE(S) OF CLINICAL INVESTIGATION TO BE CONDUCTED: ☒PHASE 1 ☐ PHASE 2 ☐ PHASE 3 ☐ OTHER _____
(Specify)

9. LIST NUMBERS OF ALL INVESTIGATIONAL NEW DRUG APPLICATIONS *(21 CFR Part 312)*, NEW DRUG OR ANTIBIOTIC APPLICATIONS *(21 CFR Part 314)*, DRUG MASTER FILES *(21 CFR Part 314 420)*, AND PRODUCT LICENSE APPLICATIONS *(21 CFR Part 601)* REFERRED TO IN THIS APPLICATION

BB-IND 009 - DMID, NIAID

MF 99999 - Vaccines of the World, Inc.

10. *IND submissions should be consecutively numbered. The initial IND should be numbered "Serial Number: 000." The next submission (e.g., amendment, report, or correspondence) should be numbered "Serial Number: 001." Subsequent submissions should be numbered consecutively in the order in which they are submitted.*	SERIAL NUMBER: 0 0 0

11. THIS SUBMISSION CONTAINS THE FOLLOWING: *(Check all that apply)*

    ☒ INITIAL INVESTIGATIONAL NEW DRUG APPLICATION (IND)   ☐ RESPONSE TO CLINICAL HOLD

PROTOCOL AMENDMENT(S):	INFORMATION AMENDMENT(S):	IND SAFETY REPORT(S):
☐ NEW PROTOCOL	☐ CHEMISTRY/MICROBIOLOGY	☐ INITIAL WRITTEN REPORT
☐ CHANGE IN PROTOCOL	☐ PHARMACOLOGY/TOXICOLOGY	☐ FOLLOW-UP TO A WRITTEN REPORT
☐ NEW INVESTIGATOR	☐ CLINICAL	

☐ RESPONSE TO FDA REQUEST FOR INFORMATION   ☐ ANNUAL REPORT   ☐ GENERAL CORRESPONDENCE

☐ REQUEST FOR REINSTATEMENT OF IND THAT IS WITHDRAWN,   ☐ OTHER _____
INACTIVATED, TERMINATED OR DISCONTINUED   (Specify)

CHECK ONLY IF APPLICABLE

JUSTIFICATION STATEMENT MUST BE SUBMITTED WITH APPLICATION FOR ANY CHECKED BELOW. REFER TO THE CITED CFR SECTION FOR FURTHER INFORMATION

☐ TREATMENT IND 21 CFR 312.35(b)   ☐ TREATMENT PROTOCOL 21 CFR 312.35(a)   ☐ CHARGE REQUEST NOTIFICATION 21 CFR 312.7(d)

FOR FDA USE ONLY

CDR/DBIND/OGD RECEIPT STAMP	DDR RECEIPT STAMP	IND NUMBER ASSIGNED:
		DIVISION ASSIGNMENT:

Figure 5.1. Form FDA 1571, Page 1.

Box 6: The FDA will assign an IND number; thus this box is left blank for the initial filing. For subsequent submissions (amendments), the correct IND number should be inserted.

Box 7: Requests information on indications; for example, "To prevent group B streptococcal disease."

12.	**CONTENTS OF APPLICATION** This application contains the following items: (check all that apply)

☒   1. Form FDA 1571 *[21 CFR 312.23(a)(1)]*
☒   2. Table of contents *[21 CFR 312.23(a)(2)]*
☒   3. Introductory statement *[21 CFR 312.23(a)(3)]*
☒   4. General investigational plan *[21 CFR 312.23(a)(3)]*
☒   5. Investigator's brochure *[21 CFR 312.23(a)(5)]*
     6. Protocol(s) *[21 CFR 312.23(a)(6)]*
         ☒ a. Study protocol(s) *[21 CFR 312.23(a)(6)]*
         ☒ b. Investigator data *[21 CFR 312.23(a)(6)(iii)(b)]* or completed Form(s) FDA 1572
         ☒ c. Facilities data *[21 CFR 312.23(a)(6)(iii)(b)]* or completed Form(s) FDA 1572
         ☒ d. Institutional Review Board data *[21 CFR 312.23(a)(6)(iii)(b)]* or completed Form(s) FDA 1572
☒   7. Chemistry, manufacturing, and control data *[21 CFR 312.23(a)(7)]*
         ☒ Environmental assessment or claim for exclusion *[21 CFR 312.23(a)(7)(iv)(e)]*
☒   8. Pharmacology and toxicology data *[21 CFR 312.23(a)(8)]*
☒   9. Previous human experience *[21 CFR 312.23(a)(9)]*
☒ 10. Additional information *[21 CFR 312.23(a)(10)]*

13.	IS ANY PART OF THE CLINICAL STUDY TO BE CONDUCTED BY A CONTRACT RESEARCH ORGANIZATION? ☐YES ☒ NO IF YES, WILL ANY SPONSOR OBLIGATIONS BE TRANSFERRED TO THE CONTRACT RESEARCH ORGANIZATION? ☐ YES ☒ NO IF YES, ATTACH A STATEMENT CONTAINING THE NAME AND ADDRESS OF THE CONTRACT RESEARCH ORGANIZATION, IDENTIFICATION OF THE CLINICAL STUDY, AND A LISTING OF THE OBLIGATIONS TRANSFERRED.
14.	NAME AND TITLE OF THE PERSON RESPONSIBLE FOR MONITORING THE CONDUCT AND PROGRESS OF THE CLINICAL INVESTIGATIONS John Doe, MD, Principal Investigator
15.	NAME(S) AND TITLE(S) OF THE PERSON(S) RESPONSIBLE FOR REVIEW AND EVALUATION OF INFORMATION RELEVANT TO THE SAFETY OF THE DRUG. John Doe, MD, Principal Investigator; Jane Doe, MD, Safety Monitor

I agree not to begin clinical investigations until 30 days after FDA's receipt of the IND unless I receive earlier notification by the FDA that the studies may begin. I also agree not to begin or continue clinical investigations covered by the IND if those studies are placed on clinical hold. I agree that an Institutional Review Board (IRB) that complies with the requirements set forth in 21 CFR Part 56 will be responsible for the initial and continuing review and approval of each of the studies in the proposed clinical investigation. I agree to conduct the investigation in accordance with all other applicable regulatory requirements.

16. NAME OF SPONSOR OR SPONSOR'S AUTHORIZED REPRESENTATIVE	17. SIGNATURE OF SPONSOR OR SPONSOR'S AUTHORIZED REPRESENTATIVE
18. ADDRESS *(Number, Street, City, State and Zip Code)*	19. TELEPHONE NUMBER *(Include Area Code)* 20. DATE

(WARNING: A willfully false statement is a criminal offense. U.S.C. Title 18, Sec. 1001.)

Public reporting burden for this collection of information is estimated to average 100 hours per response, including the time for reviewing instructions, searching existing data sources, gathering and maintaining the data needed, and completing and reviewing the collection of information. Send comments regarding this burden estimate or any other aspect of this collection of information, including suggestions for reducing this burden to:

Reports clearance officer, PHS   and to: Office of Management and Budget
Hubert H. Humphrey Building, Room 721-B   Paperwork Reduction Project (0910-0014)
200 Independence Avenue, S.W.   Washington, DC 20503
Washington, DC 20201
Attn: PRA   Please DO NOT RETURN this application to either of these addresses.

FORM FDA 1571 (12/92)   PAGE 2 OF 2

Figure 5.2. Form FDA 1571, Page 2.

Box 8: The IND application may include more than one protocol, and the appropriate phase for each protocol should be indicated.

Box 9: The information in this box informs the FDA about where to find pertinent information that has previously been submitted under an IND application or a Master File. (A Master File is a document filed with the FDA—usually, but not always—by a vaccine manufacturer. It may include a variety of information; such as facilities data, details on the manufacture of a vaccine component, or characterization of a cell line. A Master File never includes a clinical protocol). With authorization from the sponsor, the FDA can access this pertinent information to assist in the review of an IND. This activity is termed cross-referencing; a cross-reference authorization letter from the sponsor of the IND or Master File containing the pertinent information should be included with the IND filing. For example, the vaccine product prepared by an investigator/manufacturer might be a polysaccharide-tetanus toxoid (TT) conjugate vaccine. The TT may have been provided by a different manufacturer, who has information on the preparation and characterization of the TT on file at the FDA. The sponsor of the IND does not need to submit information on production of the TT as part of the IND, but the manufacturer of the TT must provide a letter to the FDA authorizing access to that information. The letter of authorization should be submitted as part of the IND application and should also be submitted to the IND or MF containing the reference information by the person granting the authorization. Even if this information is on file at the FDA, it is helpful if the submitted IND application includes the manufacturer's lot-release document, which includes a summary of the results of tests performed on the specific product being used.

Boxes 10 and 11: Self-explanatory.

Box 12: This box on the back page of form 1571, with its 10 subparts, provides a blueprint for the IND filing. All boxes that apply to the IND submission should be checked. The table of contents (subpart 2) should be clear and detailed and should indicate the pages on which pertinent information can be found. Since IND applications can be long and complex, every effort should be made to help the FDA find and refer to relevant information. The components to be included in the introductory statement and the general investigational plan (subpart 3) are detailed in the CFR. It is important to outline the clinical plans for the vaccine, because it is within this context that the FDA will review the application. For example, if the vaccine was prepared in an academic laboratory and the only plan is to undertake a Phase 1 trial before transferring rights to a pharmaceutical company, this intention should be indicated. The FDA will generally tailor its comments to the specified scenario. If the plan is not defined, the FDA may assume that the intention is to engage in large-scale trials and may ask questions about GMP production, scale-up, and lot-to-lot consistency.

INVESTIGATOR'S BROCHURE

The purpose of the investigator's brochure (subpart 5) is to describe in detail what the product is, how it is manufactured, and what is known about its safety, purity, and potency. A manufacturer conveys necessary information to clinicians in this brochure. An investigator who develops his or her own vaccine, which will be tested only under his or her supervision, is not required to submit an investigator's brochure. Institutional review boards (IRBs) often request the investigator's brochure for background information on a product. When a licensed product is being used in a study, the package insert should be included in this section. The investigator's brochure should be updated or revised as clinical and other preclinical information becomes available. In addition to the guidance in the regulations on the contents of the investigator's brochure, the document entitled "International Conference on Harmonization (ICH): Good Clinical Practice: Guideline for the Investigator's Brochure" is available from the FDA and provides helpful, current information.

PROTOCOL

The protocol (subpart 6) is the cornerstone of the IND filing and contains four basic sections: (a) the study protocol or clinical protocol; (b) investigator data; (c) facilities data; and (d) IRB data (or bioethical committee data if the study is conducted outside of the United States). The IND application may ultimately encompass several protocols.

(a) Study Protocol

A clinical protocol can be written in many styles but should always include certain elements:

- *Face sheet*—This sheet should include the title of the trial, the product to be evaluated, any associated protocol numbers, the name of the principal investigator (PI), the names of other investigators along with their institutional affiliations, and the location of all clinical sites. Investigators who are not directly part of the clinical trial but are involved in product development, assay conduct, or data analysis may be listed on the face sheet, but their roles should be clearly identified since only individuals with clinical responsibilities will be listed on form FDA 1572 and on the consent form. It is often appropriate to include nurses with major trial responsibilities on the protocol and on form FDA 1572. Protocols frequently go through several versions. It is extremely important that the protocol under consideration be identified by a date and/or a version number which should appear as a footer on each page.
- *Precis*—This section is similar to an abstract of a scientific paper and summarizes the objectives, the population to be enrolled, the design of the trial, and the outcome parameters.
- *Introduction or background*—This section should present background information in sufficient detail to permit evaluation by the FDA and the IRB of the merit of the study and how it builds on previous research. Points to be covered include the nature of the disease, with its associated morbidity and mortality; current therapeutic and preventive approaches; the relationship of the vaccine in question to this disease scenario; and clinical plans for the vaccine. It is also appropriate to discuss any controversies surrounding the

infection or its prevention. It is important to be thorough and methodical in developing the introductory section because the introductory statement and general investigational plan of the IND application (subparts 3 and 4) are frequently derived from this section. Relevant literature should be cited and the key reprints are often included in subpart 10.

- *Objectives*—The objectives of the study should be precisely worded and should not be overly ambitious—particularly in a Phase 1 study, where ascertainment of the safety of the product is the primary objective. An appropriate secondary objective might be a preliminary assessment of immunogenicity. Terms such as efficacy or effectiveness should not be used unless the sample size is sufficient to evaluate those outcomes.
- *Subjects*—This section should detail the gender, age group, and number of subjects to be studied as well as any other criteria associated with eligibility for enrollment. The latter criteria must be spelled out in detail under the headings "Inclusion Criteria" and "Exclusion Criteria." Corollaries should be stated. For example, if the vaccine product contains thimerosal as a preservative, one inclusion criterion would be no known allergy to mercury-containing compounds; and one exclusion criterion might be a known or suspected allergy or sensitivity to mercury-containing compounds.
- *Vaccines*—The vaccines to be used in the trial should be identified by name, lot number, dose concentration, and route of administration. If a control vaccine or a placebo is proposed, that should also be defined.
- *Study design*—The design of the study can often be most vividly and clearly presented by means of a schematic or a table that complements the narrative description, identifying the groups of subjects, including controls; the number of subjects in each group; and the activities that will take place at each stage of the trial (Table 5.2). For example, the timing of blood drawing, vaccine administration, and follow-up visits should all be indicated in this section. This information allows the IRB member and the IND reviewer to get a picture of the complexity of the trial. This section must also describe initial evaluation procedures and screening tests. If the study is to be randomized, the randomization procedure should be described. If any special or unusual procedures are to be performed, the credentials of the investigators undertaking those procedures should be addressed.
- *Outcome parameters*—The outcome parameters will depend on the objectives of the trial. For Phase 1 trials, the parameters will be safety and perhaps preliminary immunogenicity. Safety measures will include both local and systemic evaluations. For parenterally delivered vaccines, local safety parameters should include redness, swelling, and pain at the injection site; for intranasally administered vaccines, these parameters might include stuffiness and congestion. Wherever possible, objective measures should be provided (e.g., the size of the area of swelling or redness). Systemic measures should include fever, malaise, and nausea. Phase 1 and Phase 2 trials may also yield specimens for measurement of immunogenicity. All assays must be described in some detail so that the FDA can assess their sensitivity and specificity. For example, merely stating that ELISA measurements of IgG will be undertaken is not sufficient. Description of assays in the published

Table 5.2. Overview of the Study Design and Study Groups

Vaccines Administered and Procedures Performed: Time After First Immunization							
−7 days	0 days	1 day	2 days	3 days	1 month	2 months	3 months
S/E	D_1 B_1 NW_1	FU(T)	FU(T)	FU(V)	FU(V) NW_2	D_2 B_2 NW_3	B_3 NW_4

S/E = screen and enroll; D = vaccine dose; B = blood draw; NW = nasal wash; FU = follow-up; (T) - telephone; (V) - clinic visit

literature can be cited, but frequently detailed descriptions of standard operating procedures are often appropriate.

- *Monitoring and follow-up*—Details should be provided on the plans to monitor subjects and on the duration of follow-up, including the period during which subjects will be observed after vaccination (primarily for immediate allergic reactions). It is good clinical practice to observe subjects for approximately 30 minutes after immunization and to be prepared to take action if allergic reactions develop. In the event of a local or systemic reaction, the plans for ongoing monitoring until the reaction has resolved must be described. Details of subsequent monitoring must also be provided. For example, if responses are to be recorded by subjects in symptom diaries at defined intervals and/or if clinical personnel will contact the subjects and solicit information at specified times, these plans should be stated. Plans for longer-term follow-up should also be submitted, along with details of the information to be solicited.
- *Benefits*—The benefits to subjects in a Phase 1 or Phase 2 trial are often minimal. Investigators should be careful to not overstate the benefits of participation. If volunteers are to be compensated, the protocol and consent form should say so.
- *Risks and complications*—Investigators should strive to state potential risks accurately. For example, if pain is a risk, the word pain rather than the word tenderness should be used. When a product first enters clinical testing, the true risks are frequently unknown. Knowledge obtained from studies of similar products may be useful in anticipating risks. It is appropriate to describe precautions to be taken for the prevention and management of possible complications. If studies include minors, the risk must be minimal (equivalent to the risk of a physical exam) or only a small increment above minimal. It is wise to describe the balance of risks against clearly defined benefits of participation in the study.
- *Withdrawal from the study*—Subjects are entitled to withdraw from studies, but sometimes there are consequences of withdrawal that need to be communicated in advance. For example, a subject in a live enteric vaccine study will be shedding the vaccine strain in the stool. The subject is allowed to withdraw from the study, but may be required to remain in isolation facilities until confirmation is obtained that he/she is no longer shedding the organism. The investigator also can withdraw a patient from a study. For

example, infants who develop an immediate anaphylactic reaction or who develop encephalopathy within seven days after receipt of a pertussis-containing vaccine would not receive additional doses of vaccine. All patients who withdraw from a study should continue to be followed as specified in the protocol.

- *Stopping rules*—The protocol should also define the stopping points and criteria for terminating the study. Stopping rules may include incidence and severity of reaction to the vaccine. For example, infants who receive a live attenuated viral respiratory vaccine should not develop lower respiratory tract illness that is not explained by some other cause (such as infection with a different respiratory virus), since this may indicate inadequate attenuation of the vaccine.
- *Data analysis*—Plans and methods for the analysis of the data must be clearly defined and should be linked to the objectives of the study and to the outcome parameters. It is frequently appropriate to obtain input from a biostatistician so that statistical methods appropriate to the trial are applied.
- *Medical care*—It is helpful to delineate the extent of care to be provided by the institution and the role of any private physician both during and at completion of the formal period of study. In the event of adverse reactions requiring medical care, the level of medical care to be provided should be specifically articulated. It should be clear whether or not this care will be given free of charge.
- *Reference*—A complete but selective list of scientific literature cited in the protocol should be included. It is helpful to include a copy of all cited references.
- *Enrollment forms, case report forms, vaccine reaction diaries, and data collection forms* are part of the study protocol and must be included in the IND application.
- *Consent form*—Although a given IRB may request a certain format for consent forms, each of the fundamental requirements listed under Part 50 of CFR Title 21 must be included. Much of this information will come straight from the protocol. The protocol and the consent form must be consistent even though the consent form is written in lay language. If the protocol is modified, the consent form must reflect those changes. The consent form must be dated. One common error is omission of the statement that the FDA and the sponsor of the IND may review the subject's records. Two FDA information sheets that offer guidance on informed consent are "A Guide to Informed Consent Documents" and "Informed Consent and the Clinical Investigator." The IRB review the clinical trial advertisements as part of the information on informed consent.

(b) Investigator Data, (c) Facilities Data, and (d) Institutional Review Board Data

These data can be most efficiently provided by the completion and filing of form FDA 1572 (Figures 5.3 and 5.4). This form is protocol specific; therefore, a new protocol requires the submission of a new form FDA 1572 in the IND appli-

DEPARTMENT OF HEALTH AND HUMAN SERVICES PUBLIC HEALTH SERVICE FOOD AND DRUG ADMINISTRATION **STATEMENT OF INVESTIGATOR** *(TITLE 21, CODE OF FEDERAL REGULATIONS (CFR) Part 312)* (See instructions on reverse side)	Form Approved: OMB No. 0910-0014. Expiration Date: November 30, 1995. *See OMB Statement on Reverse* NOTE: No investigator may participate in an investigation until he/she provides the sponsor with a completed, signed Statement of Investigator, Form FDA 1572 (21 CFR 312.53(c))

1. NAME AND ADDRESS OF INVESTIGATOR

2. EDUCATION, TRAINING, AND EXPERIENCE THAT QUALIFIES THE INVESTIGATOR AS AN EXPERT IN THE CLINICAL INVESTIGATION OF THE DRUG FOR THE USE UNDER INVESTIGATION. ONE OF THE FOLLOWING IS ATTACHED.

        ☐ CURRICULUM VITAE   ☐ OTHER STATEMENT OF QUALIFICATIONS

3. NAME AND ADDRESS OF ANY MEDICAL SCHOOL, HOSPITAL, OR OTHER RESEARCH FACILITY WHERE THE CLINICAL INVESTIGATION(S) WILL BE CONDUCTED.

4. NAME AND ADDRESS OF ANY CLINICAL LABORATORY FACILITIES TO BE USED IN THE STUDY.

5. NAME AND ADDRESS OF THE INSTITUTIONAL REVIEW BOARD (IRB) THAT IS RESPONSIBLE FOR REVIEW AND APPROVAL OF THE STUDY(IES).

6. NAMES OF THE SUBINVESTIGATORS *(e.g., research fellows, residents, associates)* WHO WILL BE ASSISTING THE INVESTIGATOR IN THE CONDUCT OF THE INVESTIGATION(S).

7. NAME AND CODE NUMBER, IF ANY, OF THE PROTOCOL(S) IN THE IND FOR THE STUDY(IES) TO BE CONDUCTED BY THE INVESTIGATOR.

FORM FDA 1572 (12/92)   PREVIOUS EDITION IS OBSOLETE   PAGE 1 OF 2

Figure 5.3. Form FDA 1572, Page 1.

cation. By signing the form, the investigator agrees to conduct the study under terms articulated on the form. These are serious commitments, and violations are subject to FDA sanctions. The curriculum vitae for the PI listed on the form should be attached. There can be only one PI per form. If the PI is performing the

8. ATTACH THE FOLLOWING CLINICAL PROTOCOL INFORMATION:

☐ FOR PHASE 1 INVESTIGATIONS, A GENERAL OUTLINE OF THE PLANNED INVESTIGATION INCLUDING THE ESTIMATED DURATION OF THE STUDY AND THE MAXIMUM NUMBER OF SUBJECTS THAT WILL BE INVOLVED.

☐ FOR PHASE 2 OR 3 INVESTIGATIONS, AN OUTLINE OF THE STUDY PROTOCOL INCLUDING AN APPROXIMATION OF THE NUMBER OF SUBJECTS TO BE TREATED WITH THE DRUG AND THE NUMBER TO BE EMPLOYED AS CONTROLS, IF ANY; THE CLINICAL USES TO BE INVESTIGATED; CHARACTERISTICS OF SUBJECTS BY AGE, SEX, AND CONDITION; THE KIND OF CLINICAL OBSERVATIONS AND LABORATORY TESTS TO BE CONDUCTED; THE ESTIMATED DURATION OF THE STUDY; AND COPIES OR A DESCRIPTION OF CASE REPORT FORMS TO BE USED.

9. COMMITMENTS:

I agree to conduct the study(ies) in accordance with the relevant, current protocol(s) and will only make changes in a protocol after notifying the sponsor, except when necessary to protect the safety, rights, or welfare of subjects.

I agree to personally conduct or supervise the described investigations.

I agree to inform any patients, or any persons used as controls, that the drugs are being used for investigational purposes and I will ensure that the requirements relating to obtaining informed consent in 21 CFR Part 50 and Institutional review board (IRB) review and approval in 21 CFR Part 56 are met.

I agree to report to the sponsor adverse experiences that occur in the course of the investigation(s) in accordance with 21 CFR 312.64.

I have read and understand the information in the investigator's brochure, including the potential risks and side effects of the drug.

I agree to ensure that all associates, colleagues, and employees assisting in the conduct of the study(ies) are informed about their obligations in meeting the above commitments.

I agree to maintain adequate and accurate records in accordance with 21 CFR 312.62 and to make those records available for inspection in accordance with 21 CFR 312.68.

I will ensure that an IRB that complies with the requirements of 21 CFR Part 56 will be responsible for the initial and continuing review and approval of the clinical investigation. I also agree to promptly report to the IRB all changes in the research activity and all unanticipated problems involving risks to human subjects or others. Additionally, I will not make any changes in the research without IRB approval, except where necessary to eliminate hazards to human subjects.

INSTRUCTIONS FOR COMPLETING FORM FDA 1572
STATEMENT OF INVESTIGATOR:

1. Complete all sections. Attach a separate page if additional space is needed.

2. Attach curriculum vitae or other statement of qualifications as described in Section 2.

3. Attach protocol outline as described in Section 8.

4. Sign and date below.

5. FORWARD THE COMPLETED FORM AND ATTACHMENTS TO THE SPONSOR. The sponsor will incorporate this information along with other technical data into an Investigational New Drug Application (IND). INVESTIGATORS SHOULD NOT SEND THIS FORM DIRECTLY TO THE FOOD AND DRUG ADMINISTRATION.

10. SIGNATURE OF INVESTIGATOR	11. DATE

Public reporting burden for this collection of information is estimated to average 100 hours per response, including the time for reviewing instructions, searching existing data sources, gathering and maintaining the data needed, and completing and reviewing the collection of information. Send comments regarding this burden estimate or any other aspect of this collection of information, including suggestions for reducing this burden to:

Reports Clearance Officer, PHS   and to: Office of Management and Budget
Hubert H. Humphrey Building, Room 721-B   Paperwork Reduction Project (0910-0014)
200 Independence Avenue, S.W.   Washington, DC 20503
Washington, DC 20201
Attn: PRA
            Please DO NOT RETURN this application to either of these addresses.

FORM FDA 1572 (12/92)   PREVIOUS EDITION IS OBSOLETE   PAGE 2 OF 2

Figure 5.4. Form FDA 1572, Page 2.

study at more than one site, each site must be listed and must have an IRB approval. If the clinical trial is a multicenter trial, then the IND application must

Table 5.3. A Well-Characterized Vaccine

   I.   Chemistry, Manufacturing and Control Data
   II. Environmental Assessment or Claim for Exclusion
  III. Flow Diagrams for Growth of GBS and Polysaccharide Purification, Production
       of Conjugate Vaccine, and Bottling of Conjugate Vaccine
  IV. Vaccine Preparation
     1. Introduction
     2. Production of GBS type III capsular polysaccharide (Lot 2c)
        A. Growth of GBS type III strain M781
           1. Preparation of Columbia broth
           2. Preparation of dextrose
           3. Preparation of group B *Streptococcus* type III inoculum
           4. Growth of group B *Streptococcus* type III
           5. Harvest of group B *Streptococcus* type III
        B. Purification of type III capsular polysaccharide
           1. Purification of cell-associated polysaccharide
           2. Final purification of type III capsular polysaccharide
        C. Chemical analysis of GBS type III polysaccharide lot 2c
           1. Thiobarbituric acid assay for sialic acid
           2. Lowry assay for protein
           3. Phenol sulfuric assay for carbohydrates
           4. Spectrophotometric scan
           5. Determination of size by gel filtration chromatography
           6. Compositional analysis by pulse amperometric detection
           7. Identity
        D. NMR analysis of GBS type III polysaccharide lot 2c
        E. Composition of GBS type III polysaccharide lot 2c
     3. Oxidation of GBS type III polysaccharide
     4. Production of GBS type III polysaccharide-tetanus toxoid conjugate vaccine
        A. Purification of tetanus toxoid
        B. Conjugation of type III capsular polysaccharide to tetanus toxoid
        C. Purification of III-TT vaccine
        D. Reduction of III-TT vaccine
        E. Chemical analysis of conjugate vaccine
        F. Efficacy in mice of III-TT vaccine
     5. Composition of bulk GBS type III vaccine lot number 95-1
     6. Appendices for manufacturing protocols
     7. Protocol relating to the Lyophilization, Filling, and Safety Testing of an
        experimental type III Group B streptococcal conjugate vaccine
     8. Lot Release
     9. Permission to cross-file

---

include a signed form FDA 1572 from each PI at each site, a curriculum vitae for each PI, and an IRB approval for each site.

Both research and clinical laboratories should be listed on form FDA 1572. Laboratories that will undertake routine evaluation of clinical specimens upon which a clinical decision will be based (e.g., pregnancy testing or screening for hepatitis, or HIV infection) must be licensed/certified.

Table 5.4. A Live Viral Vaccine

  I. Chemistry, Manufacturing and Control Data
 II. Introduction
III. Synopsis
 IV. Detailed Summary
      A. Virus Seed Strain: Original Isolation, Designation and Passage History
      B. Virus Vaccine Pool Production
         1. Facilities
         2. Serially passaged AGMK cell cultures
         3. Production
         4. Pool preparation - clarification and dispensation
      C. Safety Testing Procedures and Results on the Crude, Unclarified Fluids
         1. Microbial sterility
         2. Purity (safety) in tissue cultures
         3. Animal safety
      D. Final Container Filling, Labeling, Storage and Inventory
      E. Final Container Testing
         1. Microbial sterility
         2. Reverse transcriptase assay
         3. Assay for intact tissue cell detection
         4. DNA probe analysis
         5. General safety test
         6. Virus characterization: potency/infectivity, serotype, electropherotyping
      F. Appendices
  V. Environmental Assessment or Claim for Exclusion

CHEMISTRY, MANUFACTURING, AND CONTROL DATA

Regardless of the phase of clinical investigation, information must be submitted that assures the proper identification, quality, purity, and potency of the investigational vaccine (subpart 7). However, the amount of information will vary depending on the phase of the clinical trial. The regulations in 21 CFR 312.23(a)(7)(I) and the "Guideline on the Preparation of Investigational New Drug Products (Human and Animal), March 1991" emphasize the graded nature of manufacturing and control information. To assist the FDA in its review, it is helpful to clearly define each step in the production and testing of the vaccine. Throughout the manufacturing process, it is critical that all reagents be pure, well characterized, and accurately recorded. This requirement applies to growth media as well as to reagents used in the original isolation of the strain. Standard operating procedures should always be followed, and care at every step cannot be overemphasized. To illustrate: An astrovirus was isolated in England in 1978. The virus pool was prepared with fetal calf serum from the United States, but it was not possible to document that all lots of fetal calf serum used in the isolation were from countries where bovine spongiform encephalitis is not present. Since there was no test to document that the pool did not contain the agent of this disease, purity could not be assured and the IND application was withdrawn without testing of the product in humans.

Two sample tables of contents for subpart 7 are shown in Tables 5.3 and 5.4. Table 5.3 is from an IND application for testing of a polysaccharide-protein conjugate vaccine—an example of a well-characterized vaccine. Table 5.4 is from an application for testing of a live virus vaccine, which by definition cannot be well characterized.

Requirements for testing of vaccines are found in 21 CFR Part 610, but additional requirements may also be mandated depending on the nature of the vaccine, and a listing of those tests and their requirements may or may not be found in the CFR. Testing may be thought of as belonging to the following categories: (1) tests required for all vaccines; (2) tests required for specific vaccine types, organisms, or components, including adjuvants, diluents and preservatives; and (3) tests requiring development by the manufacturer because of the unique nature of the vaccine. Part 610 "General Biological Products Standards" describes specific general safety and sterility tests required for release of all vaccines. Paragraph 610.15, "Constituent materials," states that all ingredients shall meet generally accepted standards of purity and quality and provides specific guidance on preservatives, diluents, and adjuvants; states that extraneous proteins that are allergenic (serum) must not be in the final vaccine for injection; and that only minimum concentrations of antibiotics other than penicillin can be added to the production substrate of viral vaccines. Penicillin is not allowed to be added.

The investigator-manufacturer develops and describes the tests for characterization, analysis of chemical purity and potency that are specific for the vaccine. Specifications for the ranges of acceptability for tests are also defined by the manufacturer. All of the tests and their results are provided as part of the manufacturing and safety test protocol. A lot-release document, which is a summary of the tests done, the results obtained and a description of what is a passing test, should also be prepared. It may be provided to an investigator filing a separate IND along with the permission to cross-file with the manufacturer's IND or MF for all of the information in the manufacturing and safety test protocol. It is also useful to the investigator-manufacturer and the FDA in comparing results from different lots of vaccine.

An issue of frequent concern to investigator/manufacturers is the difference between potency and stability. Potency is defined as "the specific ability or capacity of the product, as indicated by appropriate laboratory tests or by adequately controlled clinical data obtained through the administration of the product in the manner intended, to effect a given result." Stability implies that the vaccine maintains its molecular conformation and hence its biological activity for the duration of the intended storage period. The two measures are linked but distinct from each other. Potency assays measure biological activity or correlate to biological activity. These assays involve the immunization of an animal model and the demonstration of protection against challenge by the relevant organism. Stability assays must demonstrate that the vaccine remains safe and effective throughout the clinical trial period. These assays can be problematic in that they are not necessarily the same assays as those appropriate for product release. For instance, an assay that shows a vaccine to be pure and free from extraneous matter (a release criterion) may not be a suitable stability assay if its results do not correlate with potency over time. An assay for degradation products during storage may serve as part of a stability assay but needs to be correlated with a potency assay.

Early in the manufacturing plan, investigators must give careful thought to the final formulation of their vaccine. Will the product be liquid or lyophilized? If it is lyophilized, what will the diluent be? Will the formulation be for single- or multiple-dose use? Each of these decisions impacts on the testing that needs to be undertaken. For example, a lyophilized product requires a residual moisture test. A multiple-dose vaccine requires a bacteriostatic preservative. If that preservative is a mercury-containing compound, data on the amount of mercury in each dose must be provided. The type of glass and rubber stopper used may affect the product.

In addition to the chemistry, manufacturing, and control data for the investigational product, a brief general description of the composition, manufacture and control of any placebo must be submitted. A copy of all labels should be submitted in this section of the IND application or in the investigator's brochure and should be provided to each investigator. Labels must include the following statement: "Caution: New Drug—Limited by Federal (or United States) Law to Investigational Use."

Subpart 7 also requires the submission of an environmental assessment or a claim for exemption. IND applications involving vaccines frequently are exempted. An information sheet for such a claim is available from the FDA. When live agents are being studied, it must be shown that the environment will not be contaminated by the agent. Plans for studies that involve a potential environmental hazard must describe how the organism will be contained (e.g., in isolation units). For example, recipients of a live oral cholera vaccine shed live cholera bacteria in the stool. Therefore, before clinical studies were performed in outpatient volunteers, it was documented that the bacteria in the vaccine did not survive when deposited directly into the environment.

PHARMACOLOGY AND TOXICOLOGY

The information provided in subpart 8 depends on the type of vaccine involved. The CFR requires adequate data about the pharmacological and toxicological effects of the drug in laboratory animals or *in vitro* as the basis for concluding that it is reasonably safe to conduct the proposed clinical investigations. The tests that are undertaken to reach this conclusion must be tailored to the vaccine being prepared. For most vaccines the data are generated from preclinical testing. Biotechnology-derived vaccines require more classical toxicology testing.

PREVIOUS HUMAN EXPERIENCE AND ADDITIONAL INFORMATION

Any supporting data should be provided under subparts 9 and 10. The types of information requested are well described in the CFR.

COMPLETING THE IND FORM

Box 13: Most investigator-initiated IND applications will not involve a contract research organization. These organizations are used quite frequently for clinical monitoring. If any organization has been hired to carry out any of the responsibilities assigned to the investigator/sponsor of the study, they should be listed.

Boxes 14 and 15: The individual named is frequently the investigator. Depending on the study and its perceived risks, a safety monitor, internal safety committee, or data and safety monitoring board might be listed in Box 15 in addition to or instead of the investigator.

By signing the IND application, the sponsor agrees to four conditions: (1) not to begin clinical investigation until 30 days after the FDA receives the application unless earlier notification is received from the FDA that the studies may begin; (2) not to begin or continue clinical investigations covered by the application if those studies are placed on clinical hold; (3) that an IRB that complies with the FDA's regulations will be responsible for initial and continuing review of all clinical protocols filed under the application; and (4) that the investigation will be conducted in accordance with all other applicable regulatory requirements.

The original plus two copies of the entire package must be sent to the FDA at the address listed in the IND information packet. The FDA requires original photographs of data that do not copy well (i.e., polyacrylamide or agarose gels). The FDA will assign an IND number and send a letter to the sponsor acknowledging receipt of the IND application, and indicating the date of receipt. Vaccine may not be shipped across state lines until 30 days after the date of receipt of the application by the FDA unless earlier permission is received from that agency.

The FDA frequently requests additional information from the IND sponsor. This request may be conveyed by telephone and/or in writing. If the clinical trial is not placed on clinical hold, it may be initiated and the responses to the issues raised by the FDA gathered together and submitted later. This information will be filed as an amendment to the IND application. Other types of amendments are itemized in Box 11 of form FDA 1571.

The submission of an IND application is a substantive amount of work. A thorough understanding of the components and how they relate to a particular investigational vaccine should increase the likelihood of a successful outcome. This chapter provides guidance on filing an IND application, but is by no means comprehensive. Abundant information is available, and investigator/manufacturers are well advised to become familiar with the current guidelines that pertain to their vaccine type.

IND Submissions for Vaccines: Perspectives of IND Reviewers

Donna K.F. Chandler, Loris D. McVittie, and Jeanne M. Novak*

INTRODUCTION

Increasing interest has been focused on the development of vaccines and other biologics that make use of the advantages afforded by biotechnology and new immunization strategies.[1] In the United States, traditional drugs are approved by the New Drug Application (NDA) process in the Center for Drugs Evaluation and Research (CDER), of the Food and Drug Administration (FDA), while vaccines are regulated and approved through a licensing procedure prescribed by the Public Health Service Act.[2] The license applications (i.e., the Product License Application [PLA], the Establishment License Application [ELA], and the soon to be implemented Biologics License Application [BLA]) are reviewed by the Center for Biologics Evaluation and Research (CBER) of the FDA.

Clinical studies conducted in the United States to obtain the necessary safety and efficacy data to support licensure should be performed under an Investigational New Drug Application (IND).** The IND regulations, found in Title 21 of the Code of Federal Regulations (CFR), Section 312 (21 CFR 312), describe the circumstances for which an IND is required; these regulations apply to both drugs and biologics, including vaccines.

This chapter provides an overview of the IND process and the IND requirements during clinical development of a vaccine. It also offers guidance concerning the organization of information that should be included in an IND application to initiate clinical trials of investigational vaccines.

This information should serve to expedite the review process of new vaccines. Early attention to issues regarding the product and the preclinical and

---

* The views expressed in this chapter are those of the authors and do not represent official positions of the FDA or the CBER.
** The term "IND" is used to designate the documentation that is submitted to the FDA; i.e., the application and additional amendments.

clinical studies can assist in demonstration of vaccine safety and efficacy during the IND phase, which may ultimately facilitate product approval.

IND CONTENT AND FORMAT: ORIGINAL SUBMISSION

The requirements for conducting clinical trials of investigational vaccines, considered a special category of drugs, are described in 21 CFR 312. However, much of the language used in the regulations is directed toward classical therapeutic drugs. This section describes the content and format of an IND submission and outlines the information that should be included in each section of the application for an investigational vaccine.

The IND regulations in 21 CFR 312 include general provisions such as definitions (312.3), the IND content and format (312.23), administrative actions such as "clinical hold" (312.42), responsibilities of sponsors and investigators (312 Subpart D), special review of products for life-threatening illnesses (312 Subpart E), and other topics such as import and export requirements (312.110) and acceptability of foreign clinical studies (312.120).

Table 6.1 lists the sections of the IND application (Items 1–10) as prescribed in the IND regulations [21 CFR 312.23(a)]. Each new IND (original submission) as well as additional information (IND amendments) should be submitted in triplicate and accompanied by a completed, signed Form 1571 cover sheet (Item 1, Table 6.1; also see Figures 5.1 and 5.2). Form 1571 includes information about the sponsor of the IND, the names of individuals responsible for monitoring the conduct and safety of the clinical trials, the contents of the application, commitments that studies will be conducted with the approval of an institutional review board (IRB) and that the studies will not take place unless the IND is in effect, and the signature of the sponsor (or the sponsor's designated representative). A comprehensive table of contents (Item 2, Table 6.1) should be included in each original submission.

Background information should be provided and clinical and/or product development goals described in the introductory statement and general investigational plan (Item 3, Table 6.1). The introductory statement should include the rationale for the proposed investigational use of the vaccine, such as epidemiologic data (i.e., the incidence and distribution of the disease targeted by the vaccine) and expectations regarding potential product effectiveness. The specific population(s) expected to benefit from the vaccine should be described. The general investigational plan might include a description of the planned progression of clinical studies—e.g., from adults to children to infants for a pediatric vaccine, or from healthy low-risk volunteers to a high risk target population. Projections such as manufacturing scale-up or probable formulation changes should be included, so that CBER may determine whether clinical development would fall under the scope of that particular IND. For example, changes such as the incorporation of a new strain of an infectious agent or the addition of an adjuvant may require the submission of a new IND because, in effect, a new product

Table 6.1. IND Content and Format (21 CFR 312.23)

1. Cover Sheet (Form 1571)
2. Table of Contents
3. Introductory Statement and General Investigational Plan: Rationale and Background; Clinical Development Plan
4. [Reserved] (For Future Items)
5. Investigator's Brochure: Vaccine Description and Formulation; Summary of Preclinical and Clinical Safety, Immunogenicity, Activity Data; Risks and Side Effects
6. Protocol(s): (Clinical Studies)
7. Chemistry, Manufacturing, and Control Information: Vaccine Characterization, Manufacturing, and In-Process/Release Testing; Environmental Assessment
8. Pharmacology and Toxicology Information: Vaccine: Safety/Toxicity Studies (*in vitro* or *in vivo*); Immunogenicity; Activity or Efficacy in an Animal Model
9. Previous Human Experience: Reactogenicity and Immunogenicity for the Same or Similar Products
10. Additional Information

has been created. In addition, overtly different indications for the same product (e.g., use of a vaccine as both a therapeutic and a prophylactic agent in infected and uninfected populations, respectively) may require separate applications. Sponsors are advised to update the general investigational plan in the IND as clinical development proceeds. Item 4 in the IND regulations is reserved for future requirements, as needed.

The investigator's brochure (Item 5, Table 6.1) [21 CFR 312.23(a)] provides information to the investigators conducting the clinical trial. The investigator's brochure (IB) serves a purpose similar to that served by the package insert for an approved product, and it should be updated during drug development as appropriate. The IB includes a description of the vaccine and its formulation; a summary of the data (obtained from *in vitro* or animal models and from any previous clinical studies) describing safety, immunogenicity, and activity or efficacy of the vaccine; and a description of possible risks and side effects. An IB is required unless there is a single clinical investigator who is also the sponsor of the IND (sponsor-investigator, 21 CFR 312.55). For example, an academic investigator might submit an IND to examine the immunogenicity of an approved vaccine when used in a population not included in the approved indication or when used according to an unapproved schedule or regimen; or a researcher might submit an IND to investigate a vaccine that has been developed independently of a commercial manufacturer.

The clinical protocol (Item 6, Table 6.1) describes the proposed study in humans and should include characteristics of the subject population (inclusion/exclusion criteria), the vaccine dose and regimen (administration schedule), the route of administration, and the methods for monitoring safety and immunogenicity in human subjects. The content of the protocol is described in more detail below (Clinical Trials). Any study performed under IND should comply with the regulations governing informed consent (21 CFR 50, Subpart B) and IRB approval (21 CFR 56, Subpart A). The CBER usually requests a copy of the site-specific consent form.

Information on product chemistry, manufacturing, and results of quality control testing should be included in the IND (Part 7, Table 6.1). A description of the manufacturing procedures for the lot of vaccine intended for use in the clinical trial and information on product characterization are needed to support the initiation of IND studies (see Vaccine Specified Data: Manufacturing Information). Because traditional biologic products (including vaccines) are prepared from biological sources that exhibit inherent biological variability, the ability of a manufacturer to prepare a safe product consistently and reproducibly is a concern. Hence, production lots of licensed vaccines are currently subject to lot release and to specific testing as prescribed in 21 CFR 600. Lot-release testing and product characterization are discussed further below.

An environmental assessment or claim for exclusion [21 CFR 312.23(a)(7)(iv)(e)] is required in the IND submission. If, for example, the product is an inactivated vaccine, a statement explaining why the product or its use is not expected to adversely affect the environment should be sufficient to support a claim for exclusion. However, if the vaccine contains live attenuated viruses or bacteria or if the proposed clinical studies will involve challenge with live virulent organisms, a description of procedures for environmental containment and/or decontamination is needed. Some challenged volunteers may need to be isolated and/or treated with antibiotics to assure that virulent organisms are not released into the environment.

The section on preclinical pharmacology and toxicology (Item 8, Table 6.1) is an appropriate place to incorporate the results of *in vitro* and animal studies of vaccine safety and immunogenicity and models of efficacy, if available (see also Preclinical Studies).

For novel vaccine submissions, there is often no previous human experience (Item 9, Table 6.1). However, this section should contain summaries of reactogenicity and immunogenicity data obtained with the same or closely similar products, if available. Efficacy information for similar products should also be summarized. For example, safety and immunogenicity data from previous clinical studies using the product without adjuvant should be summarized in support of a proposed study of an adjuvanted product.

Item 10 (Table 6.1) should include any other relevant information that may be helpful in the review of the IND. Reprints of the critical references supporting the manufacture, testing, and use of the proposed vaccine should be included; Section 10 is a convenient location for these items. The original submission and each amendment (with additional information) should be submitted in triplicate. The pages should be numbered sequentially, with attachments and appendices included. Original photographs—rather than photocopies—of gels, blots, electron micrographs, and other analytical depictions and representations should be submitted to assist in the review process.

VACCINE SPECIFIC DATA AND INFORMATION

Overview

Product information submitted to support a vaccine study under IND is generally of two types: (1) developmental information generated with preclinical lots, and (2) specific data obtained with the lot(s) intended for clinical use. Developmental information should include data supporting the proposed formulation, such as those obtained in studies assessing the dose and regimen needed to induce an immune response in animals and efficacy studies in relevant animal models. Such developmental studies are performed as "proof of concept" and need not be repeated for each lot intended for clinical use. Specific lot-release information including potency data should be submitted to the IND for each lot prior to use of the lot in a clinical study (see Lot Release, below). Often there is a need to compare the product information for lots used in preclinical studies with that for the proposed clinical lot. Therefore, it is recommended that each vaccine lot or batch, even preclinical lots, be numbered or identified.

Manufacturing Information

A detailed flow diagram and/or a narrative description of the manufacturing process for the specific lot(s) of product intended for use in the clinical trial should be submitted in the IND. Any differences between the process used for the lots intended for clinical studies and that used for preclinical lots should be summarized.

The source and quality of starting materials, including water, should be described. If bovine or ovine materials are used in production, the source of the herds should be documented (e.g., from countries that are free of bovine spongiform encephalopathy), and information about the health of the animals should be available. Bacterial and viral seeds and master and production cell banks should be characterized. Information concerning screening procedures for adventitious agents should be included; the possibility of contamination by such agents should be considered at each processing stage, including clone selection and purification. Adequate testing should be performed on processing reagents, such as monoclonal antibodies or hyperimmune sera, to assure that adventitious agents will not be introduced into the vaccine. Required product validation and testing will depend on the type of product; for example, genetic stability should be demonstrated for recombinant constructs. Mycoplasma testing should be performed on pooled harvest fluids for products obtained from cell cultures but normally is not required for recombinant vaccines produced in bacteria or yeast or for nonrecombinant bacterial vaccines. The containers and closures used for the final product should also be described, and a copy of the labels should be included [21 CFR 312.6 and 21 CFR 312.23(a)(7)(iv)(d)]. If the vaccine is to be reconstituted with a diluent or if a placebo or other control is to be included in the clinical study, the preparation and the quality-control testing of the diluent and control should be described [(21 CFR 312.23(a)(7)(iv)(c)].

Information about the facility where the vaccine is manufactured should be provided. A flow diagram of manufacturing that also indicates where the various stages of manufacture take place is very helpful. Other products prepared in the facility (and precautions for preventing cross contamination of products) should be described. Any arrangements in which manufacturing is performed by a contract facility or shared with another manufacturer should be clearly summarized.[4,5]

Product Testing—General Standards

The IND application should contain enough information to assure the proper identification, quality, purity, and strength of the investigational product [21 CFR 312.23(a)(7)]. General standards for biologic products are described in 21 CFR 610.10-610.14. While these standards are codified requirements for licensed vaccines, it is recommended that they also be addressed for investigational vaccines to assure the safety and quality of the vaccine. For example, an injectable vaccine is expected to be sterile. The elements of product characterization and quality assurance are expected to be increasingly complete as the clinical development of the vaccine progresses.

1. Potency

Potency is defined as "the specific ability or capacity of the product, as indicated by appropriate laboratory tests or by adequately controlled clinical data obtained through the administration of the product in the manner intended, to effect a given result" [21 CFR 600.3(s)]. An adequate test for potency is needed for quantitation of the biologically active component of the vaccine and is used as one assessment of product stability. Potency testing for bacterial vaccines is discussed in a recent publication by Habig.[6]

2. General Safety

Historically, the general safety test (21 CFR 600.11), also known as the abnormal toxicity test, has been performed on the final filled product (product in final containers) to assure that deleterious substances have not been introduced during production and filling. The vaccine is injected intraperitoneally into guinea pigs and mice, and the animals are observed for normal weight gain and lack of signs of toxicity. The general safety test is usually performed for each lot of vaccine to be used in clinical trials, but it is not a substitute for comprehensive safety and toxicity testing of the product. There may be situations in which the general safety test cannot be performed as specified in 21 CFR 610.11 because of intrinsic toxicity. There are provisions for modification of the test in such circumstances, such as changing the volume or route of administration. The CBER should be consulted regarding any modification and development of the general safety test for an investigational product.

3. Sterility

Sterility testing is required for all injectable vaccines. When a vaccine is intended for oral administration, sterility, while desirable, may not be required. However, if sterility testing is not performed, assessment of the bioburden (i.e., an estimate of the bacterial and fungal load and the absence of common human pathogens, as described in the United States Pharmacopeia[7]) is usually provided.

4. Purity

Tests for purity of a vaccine may include percent residual moisture, endotoxin content, pyrogenicity, and the quantitation of residual toxic components or contaminants introduced during manufacture (for example, protease inhibitors, antifoaming agents, and organic solvents). Tests might also quantitate residual protein or DNA.

5. Identity

A test for identity on the labeled final container product is required for licensed products (21 CFR 610.14). For vaccines, an appropriate identity test might be an immunologic assay for the included antigens, such as a validated immunoblot assay or ELISA. The general appearance and labeling of the final containers should also be described so that the vaccine can be distinguished from other products manufactured in the facility.

Product Characterization

In addition to final product testing, in-process product characterization may be needed. Because biological products have traditionally been prepared from complex biological materials, reproducibility and consistency of manufacture are critical concerns. While lot-to-lot consistency is not required for Phase 1 studies, which may be initiated with a single pilot lot, product characterization should be refined during clinical development so that consistent lots that meet stated specifications can be produced. For example, in the case of polysaccharide vaccines conjugated to protein carriers, parameters such as the ratio of polysaccharide to protein, the percentage of residues on the protein carrier substituted with polysaccharide, the size of the polysaccharide substituents, the limits on the amount of free carrier protein and free polysaccharide contained in the conjugate, and the size of the conjugate may be critical to the human immune response to the vaccine. Peptide products might be characterized by SDS-PAGE, western blot, amino acid analysis, immunoelectric focusing, or immunodiffusion. Acceptance limits for these test parameters may be necessary to ensure that the vaccine can be manufactured reproducibly. Moreover, the results of such studies may need to be correlated with those of clinical studies to determine which characteristics of the product are important for immune responses in humans.

Quality-Control Testing and/or Validation

A description of the in-process tests performed for product quality and safety is needed. Appropriate quality-control testing in vaccine production might include determination of viral yields for viral products, validation of inactivation for inactivated products, amino acid analyses of peptides, validation of emulsion completeness for emulsified products, and antigen/carrier ratios for conjugates.

With any biological product there is the potential for the presence of adventitious agents. Attention should be paid to adventitious agents which might plausibly contaminate cell substrates or processing reagents, and the absence of such agents should be shown. Examples of unacceptable adventitious agents include mycoplasma, pathogenic viruses, retroviruses, parvovirus in porcine reagents, murine viruses in monoclonal antibodies, or human viruses in human sera. Likewise, bovine materials should be from herds that are free of bovine spongiform encephalopathy and pestiviruses.[8,9] Guidance documents are available to assist investigators in the development of products prepared in cell culture.[10,11]

Potentially toxic components are sometimes introduced during purification or processing of a vaccine (for example, inactivating agents such as formaldehyde, coupling reagents such as cyanogen bromide, organic solvents, affinity chromatography ligands, or media and cell components). For such substances, validation of removal or fold reduction or testing of residual levels may be needed. For vaccines that include an inactivated bacterial toxin (e.g., diphtheria and tetanus toxoids[12,13]), it will be necessary to demonstrate that the toxin does not revert to an active form on storage.

Lot Release

Currently, unless a waiver is granted, each lot of vaccine intended for sale is released by CBER prior to marketing (21 CFR 610.1). The vaccine should conform to all the applicable standards for that product; that is, it should meet the specifications prescribed in the lot-release protocol. The lot number together with the results of all tests performed should be submitted before the lot is used in clinical trials; a certificate of analysis summary is suggested. Lot-release information usually includes results of sterility, general safety, identity, purity, and potency tests. Results of in-process testing for parameters critical for the safety or manufacturing consistency of the vaccine should also be included. The lot-release summary or certificate of analysis should list the tests conducted, the acceptance criteria or acceptable limits for the test, and the results for the specific lot. Details of the test results and procedures should be attached. It is expected that lot specifications may be broad at the initiation of clinical trials, but that the specifications often will be narrowed as product and clinical development progresses.

Stability

The IND should include stability data demonstrating the integrity of the investigational product for the planned duration of the proposed clinical investigation [21 CFR 312.23(a)(7)(ii)]. Before licensure of a vaccine, it will be necessary to demonstrate the stability of the final formulation of the vaccine so that an appropriate dating period can be assigned. (Dating of product is based on real-time stability testing.) Eventually, studies will be needed to ensure that the buffers, diluents, adjuvants, preservatives, and containers and stoppers are not deleterious to the vaccine upon storage. Evaluation of immunogenicity is usually a critical part of this assessment. An adequate potency test is also important in stability studies. Depending on the vaccine, moisture content may be critical to stability. The sponsor should submit a stability testing plan to the IND.

PRECLINICAL STUDIES

Safety and Toxicity Studies

Historically, preclinical studies of vaccines in animals have not usually included acute toxicity studies like those normally performed for chemically synthesized drugs. Preclinical studies of vaccines may include (1) immunogenicity studies (which may be a measure of potency); (2) pyrogenicity studies (as part of the evaluation of vaccine purity); (3) challenge/protection studies (if appropriate animal models exist); (4) adequate attenuation for live organisms; and (5) adequate inactivation (and control for reversion to toxicity) for inactivated organisms or toxoids. For some products, *in vitro* or *in vivo* safety and/or toxicity studies may be needed. Vaccines intended for use in pregnant women may need evaluation for fetal toxicity.

Vaccines that incorporate certain adjuvants or delivery systems may require additional toxicity studies. At present, only aluminum-containing adjuvants are included in licensed vaccines. An extensive list of potential adjuvants and immune enhancers has been compiled by Vogel and Powell.[14] If the adjuvant under consideration is a novel component, acute toxicity data for the adjuvant alone will be needed; in addition, safety/toxicity data on the antigen-adjuvant formulation should be evaluated in an appropriate animal model prior to Phase 1 clinical studies. If no serious local or systemic effects are identified for the adjuvant alone, toxicity studies with the vaccine/adjuvant combination should address the potential for local inflammatory reactions, immune-mediated toxicities, and systemic toxicities.

For vaccines that contain adjuvants other than aluminum compounds, preclinical studies in animals should provide data to support the dosing levels and regimens to be used in humans as well as safety/activity profiles of the vaccine/adjuvant combination. These studies should examine the exact adjuvant/ antigen combination and formulation in a relevant animal model and should use the route of administration intended for humans. When possible, the volume and concentration administered to animals should be equivalent to or higher than the dose intended for human use. For products intended for repeated administration, the adjuvant/vaccine combination should be administered as episodic doses usually over several months rather than as daily doses over a few weeks. This "episodic" regimen will more closely mimic the regimens used for human vaccination and may reveal any potential immune-mediated toxicity. The total number of doses administered in the animal study should exceed the number of doses planned for humans. Evaluation of toxicity should ordinarily include histopathologic assessment of the injection site; complete necropsies, including organ gross pathology and histopathology; hematologic analyses; and clinical chemistries. Further considerations for the necessary preclinical studies in animals can be found in the article by Goldenthal et al.[15] Nonclinical laboratory studies that are intended to demonstrate safety of the vaccine and/or adjuvant should be performed according to Good Laboratory Practice (GLP, 21 CFR 58.1).

In addition to toxicity studies, preclinical data should be generated demonstrating the enhancing effects of the adjuvant on the immune response. It is recommended that this study use the exact antigen/adjuvant combination planned for human use and include a control group receiving the antigen(s) alone and/ or the antigens adsorbed to aluminum compounds to provide evidence that the adjuvant augments the immune response to the antigen(s). Because of factors such as special toxicity concerns or large preexisting safety databases with certain adjuvants, sponsors may wish to discuss their protocols for preclinical studies with the CBER before initiating such studies. (See also Guidance Available from the FDA, Meetings.)

Immunogenicity and Activity in an Animal Model

We recommend that the immunogenicity of an investigational vaccine be evaluated in animals (see Chapter 2). While responses in animals may not predict the exact human response, immunization of animals may yield valuable information on product safety and on the dose and regimen appropriate for clinical trials. If an animal model for the targeted disease is available, evaluation of the investigational vaccine in that model may provide preliminary evidence of efficacy.

CLINICAL TRIALS

Phases of IND studies are defined in 21 CFR 312.21 and are described by Mattheis and McInnes in this volume (Chapter 5). The initial Phase 1 trial of an investigational vaccine is traditionally an open-label study in which the assessment of safety, reactogenicity, and immunogenicity is begun in a few healthy volunteers. The information needed in the clinical protocol is delineated in detail in 21 CFR 312.23(a)(6), and a summary of the pertinent information is presented in Table 6.2.

Since the Phase 1 trial is the subject of another chapter in this volume, the discussion of clinical protocols here will be limited; however, several general recommendations can be made:

1. The specific lot(s) of vaccine that will be used in each clinical trial should be clearly identified.
2. Inclusion and exclusion criteria should be listed, the proposed screening of potential subjects should be discussed, and all clinical and laboratory monitoring to be performed should be described.
3. A copy of the subject diary or case report form that will be used to monitor local and systemic reactions should be submitted with the protocol.
4. Procedures for the immunologic assays that will be used to assess immunogenicity of the vaccine should be described.
5. Especially important for Phase 2 and Phase 3 trials is a prospective definition of endpoints and a complete description of the plan to be used for statistical analysis of study results, especially those on efficacy. The provision of specific information on these items after data analysis has been performed may affect the acceptability of the clinical data to support licensure.
6. The IRB approved consent form should be submitted.

Standardized assays are essential for evaluation of vaccine immunogenicity. Validated assays are needed to compare both the immune response elicited by different schedules and the clinical responses of various populations from study to study and from lot to lot of vaccine. Because serologic assays alone may not reflect the relevant immune response, development of the appropriate assays is especially important for vaccines that are expected to be mucosal immunogens. Furthermore, since most vaccines will need to be evaluated in trials where their efficacy will depend on accurate diagnosis of the disease, the sponsor should also develop accurate diagnostic methods and an appropriate case definition for the primary endpoint for an efficacy trial. These two measures need to be developed during the pre-IND and early IND evaluation if the candidate vaccine is to be assessed successfully.

MAINTAINING THE IND

Overview of the IND Process

When a new IND application is submitted to the FDA, an acknowledgment letter providing the IND number and listing the date of receipt of the original submission is issued to the sponsor within several weeks of receipt. A statutory

Table 6.2. Clinical Protocol Elements (21 CFR 312.23(a)(6)(iii))

(a)	Objectives and Purpose of Study
(b)	Investigator (Form 1572): Qualifications, Name and Address, Clinical Trial Site, IRB
(c)	Patient Eligibility and Exclusion Criteria; Number of Patients in Study
(d)	Study Design, Control Groups, Methods to Minimize Bias
(e)	Dose and Schedule
(f)	Monitoring to Meet Study Objectives
(g)	Monitoring for Drug Effects and to Minimize Risk

review period of 30 days begins from the data of receipt of the IND; the clinical study may not proceed before the end of this period unless the sponsor is notified [21 CFR 312.40(b)]. When the review is completed (most often toward the end of the 30-day period), the sponsor is usually informed whether the study may proceed or has been placed on clinical hold (see below). If the sponsor has not been contacted by the end of the 30-day review period, the study may proceed. For vaccines, the contact person within the CBER is usually the IND primary reviewer in the Division of Vaccines and Related Products Applications (DVRPA), Office of Vaccines Research and Review.

Clinical Hold

The grounds for which the FDA may place a proposed or ongoing study on clinical hold are described in 21 CFR 312.42; a clinical hold means that a study may not be initiated or must be discontinued (that is, no further study subjects may receive the investigational product). Also, in some situations, no additional doses of investigational vaccine may be given to subjects already enrolled in studies. For Phase 1 studies, the criteria for imposing a clinical hold include the following: (1) Human subjects are or would be exposed to an unreasonable and significant risk of illness or injury. (2) Clinical investigators are not qualified. (3) The investigator's brochure is misleading, erroneous, or materially incomplete. (4) The IND does not contain sufficient information required under 21 CFR 312.23 to assess the risks to subjects of the proposed studies. It is the impression of the authors that item (4) is the most common reason for a clinical hold of a Phase 1 study. Phase 2 and 3 studies may be placed on clinical hold for these reasons and if the trial design is inadequate to meet the stated objectives of the study (and thus is unlikely to yield complete or clearly interpretable results to support product licensure).

If a study is placed on clinical hold, the sponsor can expect to receive a letter within 30 days of being notified of the hold by telephone. This letter defines the issues responsible for the clinical hold. The trial will remain on clinical hold until responses to these issues are submitted for CBER review, the FDA determines the responses to be satisfactory, and the FDA notifies the sponsor, either by telephone or in writing, that the study may proceed.

Additionally, CBER usually sends a letter to the sponsor describing any "non-hold concerns" about the conduct or description of the proposed trial and/or the manufacture and testing of the product. These deficiencies should be addressed in a timely manner so that they do not impede subsequent clinical development (e.g., initiation of a pivotal efficacy trial).

IND Amendments

As product and clinical development progresses, new information regarding manufacturing and testing as well as plans and protocols for new or revised clinical studies should be described in amendments to the original submission. This information should be submitted with a completed copy of Form 1571, with particular attention paid to block 11 so that the submission can be precisely and comprehensively identified. A cover letter from the sponsor summarizing the content of the amendment is suggested.

Requirements for clinical protocol amendments are listed in 21 CFR 312.30. All new protocols and most protocol changes (e.g., the use of a higher vaccine dose or an accelerated regimen that may significantly affect subject safety during any study phase, changes in the design or scope of Phase 2 or 3 study, or involvement of a new clinical investigator) should be submitted for FDA review. New <u>and</u> revised protocols must also be approved by an IRB before their implementation. While it is not necessary to wait for FDA approval of these submissions (except for protocols on clinical hold), it is suggested that major protocol changes or new protocols be submitted at least several weeks before their planned implementation, allowing the FDA enough time to review and comment on the information, and thus avoiding the possibility of a clinical hold due to deficiencies discovered after the study is under way. Even more advance time is ideal for pivotal efficacy studies, which may require considerable discussion between the sponsor and the FDA. (Please note that while IRB approval must be obtained before initiating a clinical study, the protocol need not have IRB approval prior to submission for FDA review.)

Changes in product manufacture or testing and other general changes in product development plans should be reported in IND information amendments (21 CFR 312.31). Like all amendments, these should be clearly organized with regard to their intent and scope to facilitate FDA review. Before pivotal clinical trials are initiated, it is highly recommended that sponsors confirm with the CBER that the product to be used has been characterized sufficiently to support eventual licensure. As noted in IND Content and Format: Original Submission above, changes in manufacture that involve the use of a new major production component, such as a change in cell substrate or a change in or addition of a virus strain or adjuvant, usually require the submission of a new IND application.

Safety Reports

IND safety reports are described in 21 CFR 312.32. The sponsor of an IND should notify the FDA and all participating investigators in a written safety report of any adverse experience associated with use of the drug/vaccine that is both serious and unexpected. A serious adverse event is defined as "any experience that suggests a significant hazard, contraindication, side effect, or precaution" and includes any experience that is "fatal or life-threatening, is permanently disabling, requires inpatient hospitalization, or is a congenital anomaly, cancer, or overdose." Notification should be made as soon as possible and no later than 10 working days after the sponsor receives the information. The sponsor should also notify the FDA by telephone within three (3) working days of any unexpected fatal or life-threatening experience associated with the use of the drug/vaccine. *Life-threatening* in this case is considered to mean that "the patient was, in the view of the investigator, at *immediate* risk of death from the reaction as it occurred; i.e., it does not include a reaction that, had it occurred in a more serious form, might have caused death" [21 CFR 312.32(a)].

Annual Reports

An annual report is to be submitted to the IND file within 60 days of the anniversary of the date that the IND became effective. The content of the annual report is delineated in 21 CFR 312.33 and should include a summary of local and systemic adverse reactions observed in the study population as well as summary data on immune responses, if available. At the end of the study, a complete clinical report should be submitted. This report should include comprehensive summaries of adverse reactions and immunogenicity results in the subjects studied.

COMMON PITFALLS

This section summarizes some of the problems most commonly encountered in IND submissions and reiterates points made earlier regarding what comprises a clear and complete submission. Sponsors should note that "unreviewable" and incomplete submissions may result in the imposition of a clinical hold or may impede clinical development and/or licensure.

Manufacturing

Examples of problems:

1. Insufficient information is submitted to allow FDA reviewers to assess the safety of the vaccine. For example, if variable conditions of manufacture are described, the exact process used for vaccine manufacture may be unclear. In many cases, the description of in-process test results is inadequate.
2. When potentially toxic substances may be present, validation of their removal or an assay for residual components is lacking.
3. Testing for adventitious agents and/or documentation of source materials is inadequate.

The problems described above can be avoided if the sponsor submits sufficient details of the exact procedures used to manufacture the clinical lots.

Lot Information

Examples of problems: Lots intended for use in the clinical protocol are not clearly identified. In-process testing and lot-release results are not submitted. The lot-release document or certificate of analysis should include a summary table of the stage of manufacture for the test, a description of the test, the test result, and acceptance criteria. The data and individual test results should be attached to the summary document.

Preclinical Issues

Examples of problems: Vaccine IND applications have been submitted without appropriate immunogenicity data for the investigational vaccine. Frequently, experimental details of the immunogenicity and other animal or *in vitro* studies are lacking. In particular, complete information is needed on the lot, dose of vaccine, route of immunization, and assays used to evaluate the immune response. The preclinical studies are intended to support the dose to be used in the clinical trial.

Protocols

FDA protocol review is often complicated by failure of the IND sponsor to reconcile discrepancies in details of the protocol in different sections of the application or to provide enough information for the FDA to review the clinical, laboratory, and statistical validity of the methods intended for use in the study. The following points apply:

1. The protocol and other parts of the application, such as the consent form and the investigator's brochure, should be internally consistent.
2. Assays to evaluate the immune response in the clinical protocol should be described in enough detail that their utility can be assessed.
3. The subject diary and case report form that will be used to monitor reactogenicity should be submitted with the protocol and updated to reflect the most current version of the protocol.
4. Especially for efficacy trials, clearly defined endpoints and case definitions are critical parameters for acceptability of data from pivotal trials and should be submitted well in advance of the planned trial initiation.
5. Any planned interim analyses and the complete statistical analysis should be described prospectively. The statistical analysis plan should be acceptable to the FDA before the unblinding of randomized studies.

Administrative Issues

1. Three copies of each original submission and of each amendment to the IND need to be submitted to the FDA.
2. Form 1571 (the cover sheet) should be completed and signed by the sponsor for each submission (original submission and amendments).
3. When another submission (e.g., IND or Master File) is cross-referenced, the exact volume and page numbers where the cross-referenced information can be found should be cited.
4. The pages in the submission, including any attachments, should be numbered sequentially to aid the FDA in its review and communication of comments.
5. Original photographs (rather than photocopies) of gels and blots should be submitted to facilitate interpretation of the data.

GUIDANCE AVAILABLE FROM THE FDA

This chapter is offered as general guidance for the preparation of an IND for an investigational vaccine. However, the CBER views each product as unique and regulates vaccines on a case-by-case basis. For this reason, it is recommended that prospective sponsors contact the CBER for guidance early in the development of new vaccines (see 21 CFR 312.47, "Meetings"). Meetings between sponsors and the FDA prior to submission (e.g., during the pre-IND phase) may also be useful. In addition, a number of documents available upon request address regulatory, clinical, and technical issues relevant to the IND process. Finally, the comments of FDA reviewers regarding IND amendments will be conveyed throughout the development of an IND product.

Guidance Documents

FDA regulations, guidelines, recommendations, and agreements are described in 21 CFR 10.90. Guidelines "establish principles or practices of general applicability and do not include decisions or advice on particular situations.... Guidelines state procedures or standards of general applicability that are not legal requirements but are acceptable to FDA for a subject matter which falls within the laws administered by the Commissioner." In addition to guidelines, the FDA publishes a variety of recommendations that, if followed, would normally result in a product or process acceptable to the agency. These recommendations include Guidance for Industry, Points to Consider, Memoranda, and Reviewers' Guides; more recently, recommendations of the International Conference on Harmonisation (ICH) for product and clinical development have been included as guidance documents for sponsors.

A complete listing of all guidance documents and the documents themselves can be obtained by contacting the Office of Communication, Training and Manufacturers Assistance (OCTMA) at the CBER (Table 6.3). Sponsors should obtain the most recent FDA and CBER policy documents. Table 6.4 lists regulations relevant to vaccine submissions; copies may be obtained from the Superintendent of Documents, Government Printing Office, Washington, DC, 20402.

Meetings

1. Pre-IND Meetings

It is prudent to consult with the CBER for guidance early and throughout product development. Communication with FDA representatives before submission of an IND may be especially important if a manufacturer or sponsor has not previously interacted with the FDA/CBER and has no experience with IND submissions or vaccine approval or if a new product, technology, or assay is under development. If an IND is being prepared for a new vaccine, the initial contact should be with the Division of Vaccines and Related Products Applications (DVRPA) in the Office of Vaccines Research and Review, CBER. The usual initial contact would normally be with the branch chiefs of the Viral Vaccines Branch for viral or parasitic vaccines and with the Bacterial Products and Allergenics Branch for bacterial vaccines or allergenic products. The branch chiefs can provide specific guidance and recommendations. An IND packet is available from OCTMA and contains copies of selected sections of the regulations pertaining to vaccine applications, Form 1571, Form 1572 for the clinical investigator, and relevant articles and reprints.

If development of the product is sufficiently advanced, the branch chief will assign a reviewer to be responsible for further communication and to schedule a pre-IND meeting if one is requested by the sponsor. The purpose of this meeting is to discuss both product and clinical development. Before a meeting is scheduled, a summary of information can be submitted to DVRPA (approximately four weeks prior to proposed meeting dates) so that the CBER can evaluate the proposed product and initial clinical studies and can ensure that appropriate review staff will be available for the meeting. The premeeting summary materials should usually include: (1) a meeting agenda and expected list of participants; (2) a description of the product, a summary of the manufacturing process (e.g., a flowchart), a description of in-process testing, biochemical characterization, and tentative lot-release specifications for the vaccine; (3) a description of the manufacturing facility, if available; (4) a summary of preclinical data with the proposed vaccine that support a clinical study (e.g., safety studies, immunogenicity studies, neutralization assays, investigations in animal protection models); (5) previous human data for the vaccine, if available; (6) a proposed Phase 1 clinical protocol and the clinical development plan; and (7) a list of questions or issues for discussion at the meeting (e.g., formulation issues, toxicology study design, use of a novel adjuvant, and trial design).

A meeting is usually scheduled within four weeks after the receipt of adequate premeeting materials. At the meeting, it is expected that the sponsor will make a presentation, including overheads or slides, to support the planned use

Table 6.3. Availability of Guidance Documents from the CBER

- Guidance for Industry
- Guidelines
- Points to Consider
- Federal Register Notices
- ICH* Guidelines
- Blood Memoranda
- Reviewers' Guides

* ICH = International Conference on Harmonisation

A complete listing of available documents can be obtained from the FDA, Center for Biologics Evaluation and Research, Office of Communication, Training and Manufacturers Assistance and Communications (HFM-40), 1401 Rockville Pike, Rockville, MD 20852-1448; phone 301-827-2000 or 1-800-835-4709.

Documents may also be obtained by:

FAX-on-Demand: 301-827-3744 or 1-888-CBER-FAX (1-888-223-7329).
E-mail: CBER_INFO@a1.cber.fda.gov
Bounce-back e-mail index: DOC_LIST@a1.cber.fda.gov
Home page/world-wide web (WWW): http://www.fda.gov/cber

Table 6.4. Code of Federal Regulations (Title 21 CFR). Applicable to Vaccines

Part 25 - Environmental Impact Considerations
Part 50 - Protection of Human Subjects
Part 56 - Institutional Review Boards
Part 58 - GLP for Nonclinical Laboratory Studies
Part 211 - Good Manufacturing Practices
Part 312 - Investigational New Drug Application
Parts 600-680 - Biological Products Regulations

of the vaccine. Personnel that a sponsor may bring to the meeting include representatives from the regulatory affairs staff, scientific and production staff, clinical staff, and any additional consultants. Regulatory reviewers, laboratory scientists, and clinical reviewers from the FDA will normally attend.

2. Other Meetings

Once a primary IND reviewer is identified in DVRPA, this individual will be the primary point of contact for guidance and for answers to questions that arise during the IND phase. The CBER recommends meetings or conference calls whenever there are significant product-related or clinical issues that require discussion. In addition, the IND regulations prescribe meetings at particular junctures in product development (21 CFR 312.47, "Meetings"). An "End-of-Phase 2" meeting is important for a discussion of proposed Phase 3 studies and for an assessment of the status of manufacturing and testing before the initiation of a pivotal efficacy trial. A pre-PLA meeting is intended to evaluate the adequacy of the accumulated clinical data and manufacturing information and to deal with any other outstanding issues before submission of the PLA. These meetings are coordinated through the IND reviewers in DVRPA.

This chapter advises sponsors new to the IND process on how to submit an IND application and follow through with product development. These recommendations reflect the experience of the authors at the time of writing; however, changes in the regulations and recommendations over time are anticipated. Sponsors should contact the DVRPA staff for information on the most up-to-date requirements at the time they are preparing their IND application.

ACKNOWLEDGMENTS

The authors would like to thank Bette Goldman, Karen Goldenthal, Herbert Smith, Norman Baylor, and especially Van Sickler for thorough, critical reviews, as well as Paul Richman for encouragement.

REFERENCES

1. Lawrence, D.N., K.L. Goldenthal, J.W. Boslego, D.K.F. Chandler, and J.R. La Montagne. 1995. Public health implications of emerging vaccine technologies, *In: Vaccine Design: The Subunit and Adjuvant Approach*, Powell, M.F. and M.J. Newman (Eds.) Plenum Press, NY.
2. Section 351 of the Public Health Service Act (42 U.S.C. 262).
3. Code of Federal Regulations, Title 21, Part 312, Washington, DC, U.S. Government Printing Office, 1997.
4. FDA Guidance Document Concerning Use of Pilot Manufacturing Facilities for the Development and Manufacture of Biological Products. 1995. *Fed. Reg.* 60:35750–35753.
5. FDA's Policy Statement Concerning Cooperative Manufacturing Arrangements for Licensed Biologics. 1992. *Fed. Reg.* 57:55544–55546.
6. Habig, W.H. Potency testing of bacterial vaccines for human use. 1993. *Vet. Microbiol.* 37:343–351.
7. *United States Pharmacopeia*, 18th ed. 1995. Microbial Limit Tests, Section 61, Microbiological Tests, pp. 1681–1686.
8. U.S. Department of Health and Human Services. 1994. Bovine-derived materials: Agency letters to manufacturers of FDA-regulated products. *Fed. Reg.* 59:44591–44594.
9. Levings, R.L. and S.J. Wessman. 1991. Bovine viral diarrhea virus contamination of nutrient serum, cell cultures and viral vaccines. *Dev. Biol. Stand.* 75:177–181.
10. Draft of Points to Consider in the Characterization of Cell Lines Used to Produce Biologicals, CBER, July 12, 1993.
11. ICH Draft Guideline on Viral Safety Evaluation of Biotechnology Products Derived from Cell Lines of Human or Animal Origin. 1996. *Fed. Reg.* 61:21882–21891.

12. Minimum Requirements: Diphtheria Toxoid, Public Health Service, 4th Revision, March 1, 1947.
13. Minimum Requirements: Tetanus Toxoid, Public Health Service, 4th Revision, December 15, 1952.
14. Vogel, F.R. and M.F. Powell. 1995. A compendium of vaccine adjuvants and excipients, *In:* Powell, M.F. and M.J. Newman (Eds.), *Vaccine Design: The Subunit and Adjuvant Approach.* Plenum Press, NY.
15. Goldenthal, K.L., J.A. Cavagnaro, C.R. Alving, and F.R. Vogel. 1993. Safety evaluation of vaccine adjuvants: National Cooperative Vaccine Development Meeting Working Group. *AIDS Res. Human Retroviruses* 9:S47–S51.

READING LIST [Additional General References for Improving IND Submissions]

Davenport, L.W. 1995. Regulatory Considerations in Vaccine Design. *In:* Powell, M.F. and M.J. Newman (Eds.), *Vaccine Design: The Subunit and Adjuvant Approach.* Plenum Press, NY.

Durfor, C.N. and C.L. Scribner. 1992. An FDA perspective of manufacturing changes for products in human use. *Ann. NY Acad. Sci.* 665:356–363.

Goldenthal, K.L. 1992. Integrating a biological IND into a PLA/ELA: Considerations for the design and written presentation of pivotal efficacy protocols. *Regulatory Affairs* 4:277–284.

Goldenthal, K.L. and L.D. McVittie. 1997. The clinical testing of preventive vaccines. *In: Biologics Development: A Regulatory Overview,* 2nd ed. M. Mathieu (Ed.), Parexel International Corp., Waltham, MA.

Henchal, L.S., K. Midthun, and K.L Goldenthal. 1996. Selected regulatory and scientific topics for candidate rotavirus vaccine development. *J. Infect. Dis.* 174:S112–117.

Novak, J.M, J. Barrett, L.D. McVittie, and D.K.F. Chandler. 1997. The biological IND. *In: Biologics Development: A Regulatory Overview,* 2nd ed., M. Mathieu (Ed.), Parexel International Corp., Waltham, MA.

Parkman, P.D. and M.C. Hardegree. 1994. Regulation and Testing of Vaccines. *In: Vaccines,* 2nd ed., Plotkin, S.A. and E.A. Mortimer, Jr. (Eds.). W.B. Saunders Co., Philadelphia.

Scribner, C.L. 1991. Pitfalls, snares, snags, and downright awful IND submissions for biologics. *Regulatory Affairs* 3:241–248.

7 | Technology Transfer: What You Always Wanted to Know But Were Afraid to Ask

Jane A. Biddle

You hear the phrase being used and see it in print but you may still wonder exactly what technology transfer is and how it affects you and your research. This chapter will answer these questions in an understandable manner, providing scientists, researchers, and faculty members with fundamental information on technology transfer that will help in their research endeavors.

INTRODUCTION

Scientists at universities and in federal research and nonprofit laboratories are an abundant source of innovative ideas that become tomorrow's technologies. Researchers in such institutions routinely collaborate with for-profit companies, and these collaborations have generated life-saving new products as well as profits. However, before we discuss the interactions of academia, federal laboratories, and commercial organizations, we need to understand technology transfer.

What is technology transfer? Technology transfer is the translation of research results from the laboratory to the commercial sector. It is the dissemination of knowledge and intellectual property rights. Technology transfer involves various academic-federal-industry research arrangements including cooperative research agreements, sharing or exchange of equipment and personnel, research consortia, and transfer of biological research materials. In each organization involved, a particular person or an established group is responsible for managing the technology transfer program. The responsibilities of this individual or group include the management or research relationships, inventions and other proprietary material of commercial value, and administrative aspects of exchanges of research material. An important element in managing these matters is the generation of written documents that clearly state the responsibilities and roles of the parties involved. These documents generally include information on the nature of the study to be conducted, the investigators who will conduct it, the budget allotted, the protocol or research plan, the management of confidential information, the right to publish results, the ownership of proprietary information and materials, the rights to use these materials, and the responsibilities of the parties involved in the event that the study causes any harm to an individual or organization. The various types of documents are described in greater detail below ("The Mechanisms of Technology Transfer").

The overall goal of a technology transfer program at any organization is to foster productive, appropriate, and mutually beneficial research interactions and transfer of technology. More specifically, such a program aims to (1) transfer technology developed in academic and federal laboratories to the public sector for the benefit of public health in a timely manner; (2) manage the technology transfer in a way that enhances relationships with industry and increases industrial support for research; (3) provide an important service to faculty members and scientists; and (4) develop a source of discretionary income.[1]

Although the primary objectives of a technology transfer program are usually embraced by all kinds of organizations, differences between the academic and the corporate environments (Table 7.1) influence the way in which the respective organizations approach technology transfer in their policies, agreements and interactions.[2]

Despite these differences between the academic and corporate environments, successful academic-industrial relationships have been established. In these relationships, a certain balance is maintained: collaborations that promote the development of commercial products take place while conflicts of interest and commercial control of basic academic research are avoided, traditional academic freedom and independence are maintained, public interest in government-funded research is protected.

If technology transfer is to result in a commercial product, certain concepts need to be recognized by all parties: (1) basic and applied research and our understanding of this research have changed and continue to evolve; (2) the time lag from discovery to industrial application varies with the scientific field and regulatory issues surrounding that field; (3) science is more complex than we realize; and (4) product development is always subject to economic constraints.[3] It is worthwhile to keep these points in mind as you proceed with technology transfer. For your reference, terms commonly used in relation to technology transfer are defined in Appendix A.

POLICIES AND LAWS AFFECTING TECHNOLOGY TRANSFER

The United States government has passed legislation that promotes the development and accessing of new scientific technologies with the potential to enhance the quality of life and public health. These laws impact both extramural research programs (such as those funded through grants and contracts from the National Institutes of Health [NIH] to academics) and intramural research programs (i.e., those conducted at federal laboratories). Because of the beneficial effects of biomedical technology transfers on the country's competitiveness and public health, Congress has made it a national priority to bring together academia,

Table 7.1. Differences Between the Academic and the Corporate Environments[2]

Academic	Corporate
Objective: innovation	Objective: application
Science-based research	Product-focused research
Publication/collaboration desired	Patents/ownership desired
Transfer of materials	Control of materials
Government funding sought	Concerns about government's "march-in" rights, etc.
Conflict of interest	Consulting

federal research laboratories, and industry to this end.[4,5] Since 1980, federal legislation has facilitated collaborations among these entities on promising new products. Table 7.2 provides a brief history of legislation and policies related to federal technology transfer.[6] The Stevenson-Wydler Act made technology transfer a responsibility of all federal laboratories, and the Bayh-Dole Act allowed universities and small businesses to own patent rights for inventions discovered using federal funds. Between 1983 and 1985, the White House Science Council and a series of congressional hearings identified direct collaboration between federal laboratory scientists and their private-sector counterparts as a critical ingredient in the transfer of new technologies from the federal government to the private sector. Consequently, Congress passed the Federal Technology Transfer Act of 1986, which allows federal laboratories to negotiate Cooperative Research and Development Agreements (CRADAs) with industry; allows federal laboratories to assign technology, patent, and licensing rights to a collaborator before a collaboration begins; and allows inventors and their federal laboratories to share in royalties resulting from federally funded research.[7]

Federal laws allow academic institutions, nonprofit organizations, and federal laboratories the right to own and license technology derived from federally sponsored research. However, the government does have "march-in" rights (the government could require a collaborator and/or licensee to grant a license to a responsible applicant should it not commercialize the licensed technology and/or technology developed under a CRADA [35 U.S.C. § 203 and 37 CFR § 404.5]) for governmental, noncommercial purposes.[6,7] The laws emphasize that priority for licensing of technology developed with federal funds should be given to small U.S.-owned businesses and that licensing to foreign firms will be subjected to review and approval by that federal organization. In addition, Executive Order 12591 of 1987 mandated that federal research laboratories should conduct technology transfer activities, and academic and nonprofit institutions, inventions arising from federally sponsored research must report such to the federal government.[6] The National Competitiveness Act of 1989 (P.L. 101-189) allows federal contract laboratories to negotiate CRADAs and provides protection for "trade secrets" developed under those agreements. The Security Controls and Export Regulations restrict the transfer of certain high-level technologies out of the United States; this type of transfer is guided by several federal laws

Table 7.2. Brief History of Federal Laws and Policies Related to Federal Technology Transfer

---

Stevenson-Wydler Technology Innovation Act of 1980 (P.L. 96-480)
  Made technology transfer a responsibility of all federal laboratories
  Established technology transfer managers in all federal agencies
  Authorized personnel exchanges
Bayh-Dole Patent and Trademark Amendments of 1980 (P.L. 96-517)
  Authorized ownership of patent rights by universities and small businesses
White House Science Council Federal Laboratory Review Panel (1983–1985)
  Recommended formal collaboration of industry and federal laboratories
Congressional Hearings on Federal Technology Transfer (1983–1985)
  Concluded that direct collaboration between federal laboratory personnel and
    their private-sector counterparts was most important ingredient in technology
    transfer
Federal Technology Transfer Act of 1986 (P.L. 99-502)
  Allowed federal laboratories to enter into Cooperative Research and Development
    Agreements (CRADAs) with industry
  Permitted the assignment of intellectual property and license rights in advance
  Provided royalty shares to inventors and their laboratories

---

including the Export Administration Amendment Act of 1985. Despite these regulations, the transfer of most fundamental research is unrestricted.[6]

THE MECHANISMS OF TECHNOLOGY TRANSFER

The mission of a technology transfer office is to facilitate the transfer of research results to commercial firms that will develop the findings for the public's use and benefit. The translation of research results, both basic and applied, to the commercial sector is achieved by various mechanisms in the framework of either forms or agreements (Table 7.3). The type of form or agreement used depends on the scientific purpose of the action. In addition to these widely used documents, other specific documents may apply at a particular institution. Some of the documents listed are used only by government research laboratories, while others are used only by universities and nonprofit research facilities.

A point worthy of brief mention is that federal employees working in federal laboratories are currently constrained by numerous conflict-of-interest restrictions.[8] In contrast, researchers outside the federal government are subject to minimal restrictions, even if they receive federal funds. The general areas of concern related to conflicts of interest are ownership of research results from consulting; objectivity of research results, as from consulting work; use of confidential information versus the freedom to publish; research relationships with companies in which researchers or faculty have an equity position; institutions' financial interest in private companies; and use of institutions' facilities and resources by private companies.[9]

Table 7.3. Forms and Agreements Used in Technology Transfer

Invention disclosure form	Completed by any government, academic, nonprofit, or for-profit researcher who develops an invention (e.g., a therapeutic, diagnostic, or other medical device). Submitted to the institution's technology transfer office.
Confidentiality agreement	Executed between institutions before the exchange of confidential information (e.g., unpublished manuscripts). Can be one-way or two-way. Used by government, academic, nonprofit, and for-profit institutions.
Material transfer agreement (MTA)	Executed between institutions before the exchange of materials (e.g., biological materials including proprietary proteins or DNA sequences). Covers transfer receiving or providing materials. Used by government, academic, nonprofit, and for-profit institutions. Specific MTA for biologicals, while Uniform Biological Material Transfer Agreement (UBMTA) used between government, public, and nonprofit institutions.
Sponsored-research agreement	Executed between an institution and any company that wants to fund specific research projects. Typically provides for offering of commercial rights to research results in exchange for funding. Used by academic and nonprofit institutions. A similar agreement (Cooperative Research And Development Agreement or CRADA) used by government facilities.
Patent license agreement	Executed between an institution and any company that wants to make, use, and sell products developed with that institution's patented or patent-pending technology. Typically provides for up-front license fees, milestone payments, and royalties on product sales. Used by government, academic, nonprofit, and for-profit institutions.
Clinical research agreement	Executed between an institution and any company that wants to conduct drug studies at the institution's clinical facility. Typically refers to a specific protocol and provides a payment schedule. Used by institutes with clinical facilities. Also known as a clinical trial agreement or CTA.

Table 7.3. Forms and Agreements Used in Technology Transfer (Continued)

Indemnification agreement	Executed between an institution and any company that wants to conduct drug and/or research studies at that institution. Protects institution from claims relating, for example, to adverse drug reactions. Used by all institutes.
Consulting agreement	Executed between an institute and any company that wants to hire a staff member as a consultant. Used by academic and nonprofit institutes.

The technology transfer office of your institute can further explain internal and external policies on conflicts of interest that may affect you and your research. Information on the NIH policy on conflict of interest can be found at NIH's Web site (http://www.nih.gov). Academic and nonprofit institutions sometimes include information on some of their own policies on their Web sites.

MATERIAL TRANSFER AGREEMENTS (MTAs)

An MTA is executed between a researcher's institution and any organization before an exchange of materials. This type of agreement covers either the receipt or the provision of materials and is used by government, academic, nonprofit and for-profit organizations. Each organization has its own MTA, and some have two separate MTAs for the receipt and provision of materials. An example of an MTA for an academic or nonprofit institution is presented in Appendix B.

Generally, an MTA from any institution includes certain generic terms that vary to some degree with the organization. These terms address the following points: (1) The MTA is to be used for research purposes only, with no associated fee (except possibly for shipping of material(s)); (2) The use of the requested material(s) is usually restricted and is described either within the document or on an attachment; (3) Material(s) and their providers must be acknowledged in written and oral presentations; (4) There are also terms of confidentiality, publication review, and description of ownership, which could include future intellectual property rights, and licensing from such inventions and liabilities. (5) If the MTA is from a government laboratory, neither rights in intellectual property nor rights for commercial purposes are granted; and (6) will state that the federal government is held harmless from all liabilities. The Public Health Service (PHS), also requires that all materials received by their scientists originating from humans be called under 45 CFR, Protection of Human Subjects.

MTAs are generally the responsibility of the technology transfer office. However, in industry, a different group (e.g., an office of external research or legal) may be responsible. Either type of office can provide scientists with copies of MTAs for both receiving and providing. MTAs require the signatures of the scientist and the appropriate institution's representative, who is usually the director of the technology transfer office or its equivalent.

For over five years, the Association of University Technology Managers (AUTM) has been working with the NIH and other agencies and organizations to develop formats to simplify the MTA process specifically for biologicals. To this end, the Uniform Biological Material Transfer Agreement (UBMTA) has evolved. The NIH, on behalf of the PHS, has recommended that public and non-profit institutions use the UBMTA for the majority of their biological material transfers among scientists in academic and government laboratories. A simpler format, the Simple Letter Agreement, can be used for requests regarding non-proprietary materials.[10]

After institutional approval, the UBMTA is handled as a master agreement, allowing individual transfers with only an Implementing Letter. This Implementing Letter identifies the materials being transferred, the parties involved, and the terms previously agreed to by the parties in the master agreement.[11] Some biological transfers may be handled with the Simple Letter Agreement, while a few may require greater protection than is provided by the UBMTA.[11] A copy of the UBMTA as well as a further explanation of this agreement can be obtained from most technology transfer offices (including the NIH Office of Technology Transfer, telephone number, 301-496-7057; or see the NIH Web site, www.nih.gov/od/oh). The UBMTA is also included in *The AUTM Technology Transfer Practice Manual.*[11]

SPONSORED-RESEARCH AGREEMENT

A sponsored-research agreement formalizes an academic/nonprofit and industrial relationship, usually for a specific research project (Table 7.3). This type of agreement typically grants commercial rights to research results to a company in exchange for funding. The amount of funding varies, depending on such factors as the amount of time needed for the research project, the personal and internal-resources commitment of both parties, their respective involvement throughout the project, and the interest of the company. In brief, a sponsored-research agreement: (1) is funded by a company; (2) relates to research conducted at an academic/nonprofit institution; and (3) is initiated either by a company or by a principal investigator at the academic/nonprofit institution.

For successful sponsored-research relationships, the agreement needs on the one hand to incorporate the value of technological advances, promote the conversion of scientific results to commercial products, foster interactions and collaborations, and provide access for each party to the other's technology and expertise, and on the other hand to avoid commercial control of basic research, maintain academic freedom and independence, avert conflicts of interest, and protect the public interest in government-funded research. Each relationship creates its own concerns and has its own complexities. Generally, the sponsored-research agreement includes terms addressing publication, confidentiality, reporting requirements, termination, intellectual property rights, licensing of these rights developed either solely by an academic institution or jointly by academia and industry, a scope-of-research plan with timelines, and the persons responsible for various aspects of the research and the research budget, including the academic's/nonprofit's indirect costs.

A sponsored-research agreement encompasses three basic areas:

1. legal issues;
2. financial support for a particular period;
3. description of the research project, with the contributions of each party.

The technology transfer office generally prepares and negotiates the sponsored-research agreement with industry. Working with a principal investigator from its institution, the office develops a research plan that includes the title of the research project, the term of the project, the goal of the research, a detailed description of the research plan, and the respective contributions of the parties. On the basis of the research project, a budget is prepared reflecting the principal investigator's needs for the stated term of the project, such as salary (part or whole) for a postdoc and/or technician, equipment purchases, and travel (to visit the industrial collaborator and to attend meetings). The research and financial proposals are generally discussed and prepared with the collaborator's principal investigator. An open dialogue on these matters is important to the collaboration's future success. These two proposals are usually approved by the industrial collaborator before work on the legal issues is completed. Since the negotiation of legal terms takes longer, it is best to begin relevant discussions as soon as possible. Companies sometimes agree to begin the research project using an MTA and/or a letter of intent while the legal aspects of the sponsored research agreement are being completed; this decision is usually made if there is a sense from both parties that there are no major legal concerns in the agreement and that it is just a matter of time until all issues are resolved. However, if this is not the case, the research project is not begun until all legal issues are resolved. Unfortunately, this may take some time.

Once the terms of the sponsored research agreement have been completed, it is signed and dated by the appropriate person within the two organizations and sometimes by the principal investigators as well. If, as the research project progresses, both parties realize that changes in the agreement need to be made (e.g., redefining the scope of the research project or increasing financial support), an amendment is prepared, negotiated, signed, and incorporated into the original agreement.

To be mutually beneficial, a collaboration along with the agreement must:

1. maintain its science-based focus;
2. recognize the essential differences between the academic/nonprofit and the corporate environments (Table 7.1);
3. create effective scientific as well as business plans with the commercial partner;
4. support the importance and role of federal funding, (e.g., NIH grants) for basic research.[12]

The types of studies sponsored range from the evaluation of new drugs, vaccines, or assays to the performance of basic laboratory research. There is sometimes a gray area between a sponsored-research study and a material transfer agreement. Usually, if only tangible material (and not funding) is being exchanged by the company and the academic/nonprofit institution, an MTA is all that is required.

COOPERATIVE RESEARCH AND DEVELOPMENT AGREEMENT (CRADA)

Federal laboratories are staffed by scientists and engineers who collectively address virtually every area of science and technology. They have as a primary mission the transfer of federal technology and expertise to private-sector companies for commercialization to improve the U.S. economy. The CRADA is commonly used to effect this transfer. The purpose of a CRADA is to make government facilities, intellectual property, and expertise available for collaborative interactions to further develop scientific and technological knowledge into useful and marketable products.

As defined by the Federal Technology Transfer Act of 1986 (FTTA 15 U.S.C. at § 3710), a CRADA is "any agreement between one or more federal laboratories and one or more non-federal parties under which the government, through its laboratories, provides personnel, services, facilities, equipment, or resources with or without reimbursement (but not funds to non-federal parties) and the non-federal parties provide equipment, funds, personnel, services, facilities or other resources toward the conduct of specified research or development efforts which are consistent with the mission of that laboratory..." [15 U.S.C. § 3710 a(d)(1)]. A CRADA is distinct from a procurement contract or a cooperative agreement but retains the attributes of a common-law contract that binds the parties to stated terms and conditions. Generally, CRADAs are made between federal laboratories and industry in which collaboration takes place over a substantial period of time and involve government-owned inventions and laboratory-based expertise. A minority of CRADAs are made between federal laboratories and academic or nonprofit institutions. Federal laboratories may contribute staff, facilities, equipment, and supplies, but not funds. The collaborating party may contribute funds in addition to staff, facilities, equipment, and supplies. A CRADA under which the federal scientist obtains essential research material(s) that is for a one (1) year term or less and includes no other exchange of personnel or resources is conducted with a MTA-CRADA.[14]

The FTTA provides certain criteria for entering into a CRADA. The criteria for industry to participate in a CRADA include:

1. a small business firm or a consortium made up of such firms;
2. a preference for businesses that are located in the United States and that agree to manufacture products or services which embody inventions developed during the term of the CRADA substantially in the United States;[15]

3. an agreement to comply with the relevant government agency's policies and guidelines concerning, for example, research with human subjects, use of research animals, and other policy issues as they arise.

Restrictions on CRADAs include the following:

1. The CRADA may not be used to fund the services of an extramural contractor.
2. The CRADA should not place federal laboratories in competition with industry for a given technology.
3. The CRADA must confer a measurable benefit to the federal laboratory, not just to the private company.[16]
4. Through the FTTA, the CRADA grants intellectual property rights in advance to collaborators for inventions made in whole or in part by federal employees and specifies who will own the equipment purchased for the research project. The agreement also grants in advance patent licenses, assignments and options for negotiating an exclusive license agreement.
5. The private company may contribute its internal research and development funds to support the federal laboratory.

Although not generally required in choosing a CRADA partner, a competitive process is required by agencies' fair-access guidelines under limited circumstances. A notice may be published in *The Federal Register* or *Commerce Business Daily*. If appropriate, the government agency or institution may establish an ad hoc evaluation committee to review the submissions.

The CRADA model is used as the basis for all negotiations with outside parties. This model consists of five parts:

1. Legal Terminology: Introduction (outlines and defines terms)
2. Appendix A: Defines agency's Policy Statement on CRADA and Intellectual Property Licensing with Signature Page
3. Appendix B: Describes the research plan in detail. Includes statements of goals and respective contributions of the parties; abstract of the research plan for public release; notation of other related CRADAs, MTAs, and/or patent applications and patents; and Avoidance of Conflict of Interests Statement.
4. Appendix C: Details the financial and staffing contributions of the parties
5. Appendix D: Specifies company's exceptions to or modifications of the CRADA

The NIH CRADA model recently has been revised into a more streamlined and workable document. It has redefined the Legal Terminology section and Appendix A of the above-mentioned model and combined them into one compact section. In the NIH CRADA model, the Research Plan is now denoted as Appendix A, the Financial and Staffing Contributions as Appendix B, and the Exceptions or Modifications as Appendix C.

The FTTA provides for a 30-day period in which a CRADA can be disapproved or modified after its finalization by an agency or institution. When no changes are required, the CRADA is signed. The date of the last signature may be specified as the starting date for the CRADA. There is no mandatory term length for a CRADA but a 4- to 5-year term is often designated and can be extended by the mutual agreement of the parties if there is no substantial change in the research plan. In order to expedite the commencement of the research prior to final execution of the CRADA, some agencies sign an interim letter of intent with the collaborator.

Several aspects of the CRADA benefit private industry:[17]

1. Unlike procurement contracts, cooperative agreements, and grants, CRADAs are not subject to the Federal Acquisition Regulations. A CRADA need not be opened to competitive bidding, need not grant an exclusive license to intellectual property, and need not be advertised, as is required for federal licenses obtained from government-owned and government-operated facilities (GOGOs).
2. The federal laboratory involved in the CRADA usually contributes at least 50 percent of the development costs to the research project. This type of collaboration represents an excellent means for industry to leverage its resources in research and development of new products and processes.
3. There must be at least one industrial partner for a federal laboratory in a CRADA. In certain situations, U.S.-based subsidiaries of foreign-owned companies may be eligible.
4. Under the Freedom of Information Act, federal laboratories may protect from public access, for up to five years, information produced under a CRADA and designated as commercially valuable. Likewise, trade secrets and other confidential information obtained from a nonfederal partner during research or as a result of activities covered under a CRADA need not be disclosed.
5. The ownership and disposition of any intellectual property used for or developed from a CRADA are specified in the agreement's policy statement. Results created solely by a private collaborator's employees or jointly with federal employees under a CRADA may be copyrighted or patented and owned by collaborator. The rights to a jointly developed invention are owned jointly by the private partner and the government. However, the government can grant to its collaborator in advance a license, an assignment, or an option to license any intellectual property produced under the CRADA by federal employees or their contractors. This policy is a major departure from other types of federal contracts. Intellectual property rights to an invention made solely by the collaborator are owned by the company, while rights to an invention made solely by federal workers are owned by the government.
6. Recently, Congress passed legislation [Public Law 104-113, The National Technology Transfer and Advancement Act of 1995] requiring government facilities to assign the title to any intellectual property resulting from a

CRADA to the private-sector collaborator. A summary of this bill is presented in Appendix C.[18] In exchange, the government is entitled to "reasonable compensation" and would retain "march-in rights," through which the government could require the partner to grant a license to a responsible applicant should it not commercialize the technology developed under the CRADA. The government always retains an irrevocable, royalty-free, worldwide, nonexclusive license to practice on its behalf a federally owned invention developed under a CRADA.

A CRADA can be initiated by government and/or industry. It is usually the private sector that contacts the federal laboratory after identifying the technology of interest. As mentioned above, technologies available within federal laboratories for collaborations can be located by reviewing the current periodicals in the field of interest as well as through the various government information centers, such as the National Technical Information Service (NTIS) operated by the Department of Commerce; the Energy Sciences and Technology Software Center operated by the Department of Energy; the Office of Technology Transfer operated by the National Institutes of Health, which covers the Food and Drug Administration (FDA); the technology transfer office of each of the individual institutes that make up the NIH; and the Office of Technology Transfer of the Centers for Disease Control and Prevention (CDC). Within the U.S. Public Health Service, responsibility for CRADAs, including negotiations and authority, lies with the respective technology transfer offices of the FDA, the CDC, and the individual institutes of the NIH. Another source of information on government technologies available for collaboration are publications such as *Research and Development Magazine, The Federal Register,* and *The Commerce Business Daily,* and the respective Web sites of the various departments, agencies, institutes and centers.

Although the principles of the CRADA are the same for all federal laboratories, each federal agency has its own CRADA forms and procedures, corresponding to its mission. For further information regarding the various agencies, regulations and CRADA models, contact the respective organization's Web site.

Recent increases in CRADA activity reflect the response of industry and other institutions to the government's invitation to combine their resources with federal technology in order to achieve a competitive advantage in today's world markets. CRADAs are expected to further the federal objective of improving the U.S. economy while enhancing public health.

INVENTIONS

Before discussing patents, we need to understand the process leading to the granting of a patent. An *invention* is a new technical development or discovery within a particular scientific field; it may or may not be patentable. The invention recognition is the important step in the invention-based technology transfer process. The conception of an invention is the arrival at a specific solution for a problem by an inventor. The "actual reduction to practice" of an invention is the completion and testing of an embodiment of the invention under certain conditions.[19] The investigator of this invention needs to stop and ask: "Is the discovery new? Is the discovery useful?" If the responses to these questions are positive, then disclosure to the institutional administrator responsible for handling inventions or intellectual property is necessary. A scientist should make this assessment each time an experiment is completed or as the data are collected and during the preparation of manuscripts, presentations, and grant proposals. Even after public disclosure of a discovery, in the form of an oral presentation, an article, a poster session, or a thesis, an invention should still be reported to technology transfer personnel.

An invention disclosure is a written document that describes an invention. This report should include sufficient detail to permit an evaluation of: (1) the scientific and technical merit of the invention, with related articles and/or manuscripts attached; (2) whether and how the invention can be protected; (3) whether the work that led to the invention was supported by government or private funds; and (4) the invention's commercial value.

As a condition of employment at an institution and in line with the institution's patent policy, inventors usually have to assign their interest in any inventions they develop at the institution while employed by the institution to the institution, which becomes the sole owner of the inventions. In return, the institution files, prosecutes, and supports patent activities on the invention. Both the institution and the inventors benefit from any royalties obtained from the licensing of the invention. The patent policies of many but not all institutions stipulate that a specific share of any resulting income shall be returned as personal income to the inventors. The assignment requirement applies to federal government inventors, and the National Technology Transfer and Advancement Act of 1995 (see Appendix C) stipulates the maximum royalty income a federal inventor can receive in a year. It is important to note that each federal agency has its own patent and royalty distribution policies.

INVENTORSHIP

It is important to establish and resolve the issue of who is an inventor before filing a patent application. As noted in the U.S. Constitution and based on U.S. patent law, a patent is a personal entitlement of the legal inventor(s).[20] A patent may lawfully be issued only in the name of those who meet the criteria for inventorship. Failure to identify the legal inventors correctly on a patent application is regarded as fraud against the U.S. Patent and Trademark Office. Incorrectly naming of an inventor on a patent application must be corrected since such an error can serve as a basis for invalidating the patent.

Under U.S. patent laws and regulations, inventorship has a strict legal meaning. The law specifies that only those who have made an independent conceptual contribution to an invention are legal inventors in the United States.

(Inventorship may be interpreted differently in foreign countries.) The criteria for inventorship are as follows:[21]

1. An inventor is one who, alone or with others, first invents a new and useful process, machine, composition of matter, or other patentable subject matter.
2. If an invention involves more than one inventor, it is known as a joint invention, with multiple inventors known as joint inventors, or coinventors.
3. Inventorship and authorship are not the same. All coinventors are coauthors of a publication describing an invention, while coauthors may not necessarily be coinventors.
4. The most important criterion for inventorship is probably the initial conception of the invention. The courts have ruled that, unless a person contributes to the conception of the invention, that person is not an inventor.

A person is an inventor only if his or her intellectual contribution renders the invention complete and operative as it is to be applied in practice. If you eliminate the contribution and the invention still functions in the manner conceived and reduced to practice, then the person who made the contribution is not considered an inventor. According to 35 U.S.C. § 116:

"When an invention is made by two or more persons jointly, such persons shall apply for patent jointly and each [sign the application and] make the required oath, except as otherwise provided in this title. Inventors may apply for a patent jointly even though (i) they did not physically work together or at the same time, (ii) each did not make the same type or amount of contribution, or (iii) each did not make contribution to the subject matter of every claim of the patent."

Legal determination of inventorship is made by a patent attorney or a patent agent in relation to patent claims. The test of inventorship is whether a person has made an original, conceptual contribution to one of the claims of the patent. To be a sole inventor, a person must be responsible for the conception of the invention as described in all the patent claims.

A good reference book of organizational listings to help inventors in government, university, and corporate research laboratories to market their inventions is *The Inventor's Desktop Companion* by Richard C. Levy, (444 pp., Visible Ink Press, Detroit, Michigan).

PATENTS

A patent is a contract created by law between the inventor and/or owner and the government. It is considered personal property. Three types of patent are granted in the United States: utility, design, and plant. When the owner of an invention deemed patentable under established U.S. patent law discloses it and publishes a description of it, the federal government grants the owner (also known as patentee), the property right to exclude all others from using the invention for a specified number of years (14 for design; 20 from filing date of the earliest application pertaining to the invention for utility and plants see below). After this term, the patent then expires, the property right ceases, and the invention enters the public domain. For example, a patent covering a product or a process entitles the patent owner to prevent anyone else from making, using or selling that product or process for a certain period. U.S. patent law requires that the invention, once granted as a patent, be made public. Thus, by reading the patent and obtaining any relevant materials required by the U.S. Patent Office (or other patent offices worldwide), such as deposits of biological materials claimed in the patent, others are able to practice the invention if given a license (or some other form of legal permission) to do so by the patent owner.[22]

The owner of a patent has the right to prevent others from practicing the invention covered by a patent.[23] However, the owner has no absolute right to practice the invention. For example, the practice of a patented invention may infringe on a dominating patent owned by another party and would require licensing by the holder of this other patent. Moreover, the necessity for obtaining approval from a federal agency (e.g., the FDA) to sell a patented pharmaceutical and biological product or medical device reflects the lack of absolute right of a patent owner to practice an invention.[22]

An invention may or may not be patentable. What does this mean? An invention is eligible for a utility patent under U.S. patent law if it meets the specified criteria: i.e., if it is novel, unobvious, useful and enabling. Three of these criteria—novel, unobvious, and enabling—are also necessary for plant and design patents. Under U.S. patent law, "Whoever invents or discovers any new and useful process, machine, manufacture, or composition of matter, or any new and useful improvement thereof, may obtain a patent..." [35 U.S.C. § 101]. U.S. patent law further defines process as "process, art or method, and includes a new use of a known process, machine, manufacture, composition of matter or materials, and improvements to any" [35 U.S.C. § 100]. Some examples of inventions that may be patentable are new compounds, new uses of known compounds, new mechanical devices, new instruments, new chemical processes, new methods of making and using genetically engineered products as well as the products themselves, new plants, new computer systems and software, and new life forms. Patents can be obtained only on a tangible embodiment of an idea; in other words, simple ideas, ways of doing business, or mental processes cannot be patented as well as products of nature. Biotechnology inventions that can be protected with a utility patent include the following exemplary products and processes: non-naturally occurring DNA, RNA, vectors, transformed hosts, cell lines including hybridomas, foods, apparatuses, non-naturally occurring proteins, enzymes, receptors, hormones, growth factors, lymphokines, monoclonal antibodies, vaccines, animals, plants, fermentation, cloning, expression, microorganism cultures, food preparation, plant and animal husbandry, immunization methods, diagnostic assays, and purification procedures.

The criteria for patentability are defined as follows:

Novel:

The invention must be new; i.e., different from what has been done or known previously in a particular field of study. New uses of known processes, machines, compositions of matter, and materials are patentable.[24]

Useful:

An invention must perform a useful function or satisfy a need. Its utility does not have to be of commercial value, nor must it be superior to other means of achieving the same or a similar purpose. Nevertheless, documentation of superiority to existing products or processes may help to prove that an invention is not obvious to those skilled in the field at the time of its creation.[25]

Unobvious:

An invention must be unobvious to persons with an ordinary degree of skill in the particular field at the time the invention was made. Art or prior art includes patents, publications, and general knowledge in the field(s) of science that the invention covers as they existed at the time of discovery.[26]

Enabling:

An invention must be described in a full, clear, and concise manner in a written patent application. The application must include at least one claim particularly pointing out the scope of protection for the invention sought by the inventor(s). In addition, the description of the invention in the application must enable a person of skill in the relevant field to both make and use the invention. The application must disclose the best mode of practicing the invention known to the inventor(s) at the time of filing.[27]

The novelty requirement ensures that a patent is not granted when the claimed invention is identical to an invention found in the prior art. The unobviousness standard ensures that an invention, even though novel, is not granted patent protection if it would have been obvious at the time it was made to a person with ordinary skill in the art or technology to which it pertains.

To review the process so far: You have stopped and reflected on your work, and you believe that you have a discovery that may or may not be patentable. You contemplate whether your invention satisfies the criteria of patentability, and you conclude that it does. You contact your technology transfer office to obtain an invention disclosure report, complete it, and return it to the technology transfer office. It is here, usually, that it is determined whether your invention is indeed patentable and whether it is worthwhile for your institution to file a patent application (i.e., whether the institution will be able to license your invention and recoup its cost). Each institution's technology transfer office has its own procedure for determining whether or not to file a patent application. Some offices use outside organizations to advise them on the patentability and marketability of their inventions. Others use in-house technology transfer staff or establish a technology transfer committee composed of scientists, patent attorneys/advisors, and business/marketing representative who meet regularly and evaluate all inventions.

Your institution's technology transfer office decides to file a patent application for your invention. What does this mean? A patent application is similar to an extensive review article that is submitted on behalf of the inventor(s) to the U.S. Patent and Trademark Office (PTO) in Washington, DC. The form and contents of the application are largely determined by requirements set forth in the patent statute and in the PTO rules promulgated therein. Basically, a patent application consists of four parts:[28]

1. a specification that identifies and describes the invention in detail;
2. a drawing, when necessary;
3. a claim, the last section that precisely defines those elements that render the invention novel and patentable; and
4. an oath by the inventor(s) of the invention, as specified by law.

In addition, a filing fee for the patent application must be paid.

The specification is a narrative presentation that describes in broad terms the background of the germane previous knowledge or "prior art." The specification enables a person of ordinary skill in the art or field relevant to the invention to practice the invention. The specification must include the best mode of practicing the invention known by the inventor(s).[27,29] If an invention requires the use of a biological material; if access to that material is needed to satisfy the statutory requirements for a complete description, for enablement, and for setting forth of the best mode of practice, and if the material is not known or readily available and words alone cannot sufficiently describe how to make and/or use the material in a reproducible manner, the biological material can be provided through a deposit at a recognized International Depository Authority (IDA) (e.g., the American Type Culture Collection [ATCC]).[30]

Each claim represents one or more of the essential conceptual elements that make up the invention. As a whole, the claims define the scope of the invention by describing the specific features that distinguish the invention from the prior art.[31] The claims also provide the basis for legal enforcement of the patent. Just as your house is a tangible piece of property that has boundaries, so does a patent by its claims. When someone "trespasses" on your patented property without a license, they do so illegally and are subject to the charge of patent infringement.

The PTO has routinely issued patents for medical procedures. However, Congress recently passed the Omnibus Appropriations Bill, which restricts patent protection for medical procedures such as surgical methods. This law, (P.L. 104-208) amends 35 U.S.C. § 287 by adding that a medical practitioner's performance of medical activity which would constitute an infringement under 35 U.S.C. §

281, § 283, § 284, and § 285, shall not apply against the medical practitioner or against a related health care entity with respect to such medical activity. In the statute, a medical activity is defined as "the performance of a medical or surgical procedure on a body." This definition of medical activity, however, does not include the use of a patented machine, manufacture or composition of matter in violation of a patent; nor does it encompass the implementation of a patented method of use of a composition of matter in violation of a method patent. If the statute does not apply to an activity, a medical practitioner will still be subject to infringement liability for undertaking the activity. A patent claim to that activity will be unaffected by the statute. Thus, the law now exempts physicians, other licensed health professionals and any entity these individuals are affiliated with from infringement liability for "medical activities."[32]

Patent applications are prepared by patent attorneys or agents. However, the inventor(s) may prepare and prosecute their own application. A patent agent is not an attorney but has a technical background and is registered to practice before the PTO.

The patent application constitutes a "constructive reduction to practice" of the invention described within, which may provide the basis for a claim that the inventor is entitled to priority over other inventors of the same invention. It also serves as the basis for foreign patent applications.

THE PATENT PROCESS

When the patent application has been completed, it is filed with the PTO, which assigns it to the examining group in charge of the classes of inventions to which patent application relates. The application is next assigned to an examiner in the group, who processes it in the order in which it was received or in accordance with examining procedures established by the PTO.[33]

Every application filed in the PTO receives a filing date, which is the date the patent application was received by the PTO, and a serial number. The serial numbers are used for identification purposes by the PTO and are given in chronological order.

The examination of the application consists of a study of its compliance with the legal requirements for patentability and a search of the prior art, including prior U.S. patents, foreign patents, and the literature. After the examiner has completed this review for compliance, the inventor(s), along with their represented patent counsel, are notified by mail of the examiner's decision, known as an "office action." If the invention is not considered patentable, the claims will be rejected. Some or all claims are usually rejected on the first of three possible actions by the examiner. Only rarely is an application allowed as filed.[33]

Through their patent counsel, inventor(s) respond to the office action in writing within a given period (no longer than six months from the date of mailing of the office action). The inventor(s) and the patent counsel must respond to every point of objection and rejection. Their response must clearly be a bona fide attempt to advance the case to final action. If, after a review of the examiner's action, the application must be amended in order to overcome a rejection, then the patent counsel (with inventor's advice) must clearly point out the patentable novelty of the revised claims presented. On receipt of the response, the examiner will review and reconsider the application. As after the first examination, the patent counsel and inventor will be notified of the second office action with any rejections or objections. This second office action can be final, and a third office action is always final. A response to a final office action is limited to an appeal in the case of rejection of any claim, while any further amendment is restricted to a petition to the Commissioner of the PTO but only with regard to objections or requirements that do not concern the rejection of any claim. Thus, a response to a final office action must include either cancellation of claim(s) or an appeal on the rejection of each rejected claim and, if any claim is allowed, compliance with any requirement of objections to its form.[33]

Interviews with examiners may be arranged to discuss office actions. However, an interview does not eliminate the requirement for a response within the given period.[33]

A common problem cited in the initial review by the examiner is that a single application claims two or more inventions. By law, the PTO can issue a patent on only one invention described in a single application. Thus the application must be restricted to one invention. Other inventions may be made the subject matter of separate applications, which, if filed while the first application is still pending, will be entitled to the benefit of the filing date of the first application. This restriction requirement is made by the examiner before further action is taken.[34]

If an application is found to be allowable by the examiner, a notice of allowance will be sent to patent counsel and inventors. A fee for issuing the patent is due within three months from the date of notice. On the date the patent is granted, the patent and the record of patent, including the prosecution proceedings, become available to the public. All utility patents granted on or after December 12, 1980, are subject to maintenance fees, due 3-1/2, 7-1/2, and 11-1/2 years from the date the patent is granted.[35]

The U.S. PTO is a fee-for-service organization within the Department of Commerce. Thus, almost all interactions with the PTO entail payment of a fee. The filing fee for patent applications (except for design applications) consists of a basic fee and additional fees. The Commissioner of the PTO has the legislative authority to raise all fees involved in the patent prosecution process in the PTO without a prior express legislative act. Fees usually increase every October; the figures are published in *The Federal Register* and are available from the PTO. The PTO fees are set by 35 U.S.C. § 41.

The PTO has two pay scales: a small-entity scale (for independent inventors and nonprofit organizations) and a large-entity scale. The definitions of these entities are set forth by the PTO and by the Small Business Administration and are found in 37 CFR § 1.9(c), (e) and (f) (PTO Rules) and 13 CFR § 121.3–18 (Small Business Administration Rules). The small-entity fee is one-half of the large-entity fee.

A more detailed explanation of the U.S. and foreign filing and patent processes can be found in *The AUTM Technology Transfer Practice Manual* (Volumes I–III), *The AUTM Educational Series: An Inventor's Guide to Patents and Patenting* by Lisa von Bargen Mueller (Number 1: 1995) [The Association of University Technology Managers, Inc.: 1995], and *The Licensing Executive Society's (LES) Technology Transfer Manuals.* In addition to these text sources, the following Web sites may be of help:

www.ladas.com
An overview of U.S. patent practice in biotechnology inventions

www.foleylarnder.com
Legal and scientific information relevant to biotechnologists from the law firm of Foley & Larnder. Also has a European Intellectual Property law page and a listing of their multidisciplinary attorneys.

PROVISIONAL APPLICATION

An important change in U.S. patent law occurred after June 8, 1995, and resulted from the General Agreement on Tariffs and Trade (GATT). This change made it possible to file a simplified, less expensive provisional patent application. The provisional application preserves rights to the invention for up to one year, at the end of which a standard patent application must be filed if the inventor wishes to obtain a patent; or otherwise it becomes abandoned.[36] The provisional filing date may also be used as a priority date for foreign filing.

A provisional application is an informal document that does not require claims, oaths, or formal filing papers. Rather it contains only a specification and, (in some cases) a drawing. Conceivably, it could be a do-it-yourself project. However, the specification must satisfy the requirements of description, enablement and best mode. In many cases, institutions file the inventor's journal articles as provisional applications in order to preserve their patent rights, including foreign rights (if filed before publication).

The provisional application is not examined and cannot become a patent. One benefit of a provisional application is that its filing does not cause the clock to start running on the 20-year patent term. Thus, by the filing of a provisional application, it is possible to get a term that is 21 years from the filing date instead of 20 years.[37]

It is important to keep in mind that the provisional application protects only what is disclosed in the submitted application or articles. It does not cover any improvements that are made in the interval before the formal patent application is filed.

The filing cost for a provisional application is $75 for small-entity or nonprofit institutions and $150 for large-entity institutions. These filing fees are significantly lower than those charged for regular or standard patent applications.

EFFECTS OF PUBLIC DISCLOSURES ON PATENTS

Disclosure can be defined as a nonconfidential release of the critical aspects of an invention by written or oral description, by use, or by any other means.[38] Simply displaying the invention so that its critical features are readily discernible is considered a disclosure, as is the distribution of samples of the invention (such as a compound) whose critical features can be discovered by analysis—even if the analysis never actually takes place or is conducted without your knowledge.[38]

What is an enabling disclosure? To be considered enabling, a disclosure—written or oral—must describe an invention in sufficient detail and with sufficient specificity that a person of ordinary skill in that art or field of science at that time can make and practice the invention without an unreasonable amount of experimentation.[39,40] To be enabling, the divulgence does not need to describe information in such detail that the invention would be obvious to such a person at that time. In the case of a new chemical compound described by a name or a structural formula, the disclosure is not enabling unless a method of making the compound either is described in a publication or an oral presentation or is obvious to a person skilled in that art at that time. However, subsequent disclosure of the method of making the compound can render the originally nonenabling disclosure enabling as of the date of the subsequent divulgence.[39]

If you disclose your invention either by publishing or by presenting at a meeting prior to filing a provisional application or a standard patent application, foreign filing rights may be forfeited.[41] In the United States, a one-year grace period follows the release of an enabling disclosure prior to the filing of an application. In the rest of the world, the patent laws are based on absolute novelty and thus preclude the filing of a patent application following public disclosure. In other words, patent laws in the United States grant a patent to the inventor who first discovers a novel product or process, and the system recognizes the fact that inventors do not always realize that they must protect themselves before disclosing their inventions. In almost every country other than the United States (and in some cases Canada), patent laws award a patent to whoever is first to file with absolute novelty.[42] An inventor forfeits rights to a U.S. patent if an enabling disclosure is made either in print or orally more than one year before a U.S. provisional or patent application is filed.[41] Foreign filing rights are also lost at the time of the divulgence. Thus, disclosure must not precede the filing of a provisional or standard patent application. In summary:

IF YOU PUBLISH OR PUBLICLY DISCLOSE **BEFORE** FILING AN APPLICATION:	**PUBLIC DISCLOSURE DATE**	IF YOU PUBLISH OR PUBLICLY DISCLOSE **AFTER** FILING AN APPLICATION:
• Foreign rights are lost.		Both U.S. and foreign rights are retained.
• U.S. rights are retained IF you file within 1 year.		Foreign filings must take place within 1 year.

How do you preserve your patent rights and still publish or talk about your research? Talk with your technology transfer personnel and discuss how best to proceed. For example, given the timing of the intended disclosure, is there enough time to prepare an invention disclosure report, to conduct an internal review, to undertake discussions with patent counsel, and so on, or do time constraints make it desirable to file a provisional application and thus to secure a priority date? Ideally, you will notify technology transfer personnel of a possibly patentable invention when you are preparing a manuscript describing it or when you have just submitted the manuscript for review.

You can publish and protect the commercially valuable parts of your research by:

1. planning ahead;
2. coordinating public disclosure with patent filing;
3. preparing and submitting a disclosure of invention to the technology transfer office as early as possible.

Keep in mind that patent protection is complementary to publishing/talking and is essential to the successful commercialization of many inventions.[43]

RECORD KEEPING

The careful recording of ideas and laboratory results is a matter of routine for commercial researchers. This record keeping is a legal documentation of the day's work; each lab notebook entry is completed and up-to-date, signed and witnessed. In contrast, in academic, nonprofit, and federal government laboratories, record keeping is not always routine, and irregular working hours often make it hard to find suitable witnesses. Despite such difficulties, the routine maintenance of a witnessed laboratory notebook is important and should be attempted. Such documentation could serve as valuable repositories of new ideas.[44]

The ideal practice is to use bound notebooks for records, making entries on a daily basis on consecutively numbered pages. This diary-like format provides a day-to-day chronology. The notebook should also be used to record research- and invention-related ideas, laboratory data, drawings and computer printouts. If possible, each entry should have a title and should be continued on successive pages. The recording should be done in ink, and mistakes should be indicated by a line through the text or drawing to be deleted, with the entry of correct information thereafter. Recordings should not be erased or whited out. Separate pages and photographs should be pasted to notebook pages and referred to in an entry. Materials that cannot be incorporated into the notebook should be referenced by an entry. All entries should be signed and dated both by the researcher at the time of recording and by a witness, either at the same time or within a reasonable period. The witness should be capable of understanding notebook entries but should have nothing to do with the work from which the entries result. Usually, an inventor and coinventor cannot serve as their own witnesses. It may be best to set aside a time for making notebook entries and to arrange to have two or more colleagues serve as witnesses on a regular basis.[44]

Notebook keeping is a routine process that can take little time and effort and yet can become an invaluable asset to research in progress and ultimately may protect the rights to which the inventor is entitled.

INVENTION RIGHTS

Most government grant and contract awards for the performance of experimental, developmental, or research work incorporate standard patent-rights clauses, which state that, subject to certain limitations, the rights to any invention are usually owned by the contractor or grantee organization, in accordance with the Bayh-Dole Act (Public Law 96-517).[45] However, the organization must elect in writing whether or not to retain title to the invention during the next one or two years after the required disclosure—for example, to the Extramural Invention Reports Office at the NIH. Any organization that elects to retain title is obligated to file an initial patent application within a reasonable period: one year from either a public disclosure date or the filing date of a provisional application or prior to any statutory bar date (e.g., foreign patent application). Thus, the government strongly discourages organizations from attempting to protect or license inventions as trade secrets without filing for patent protection.[45,46]

If the organization elects not to file for a patent, it must so inform the government agency where the grant or contract originated, which then has the right to take title. (The title does not flow to the inventor by default.) That particular agency's staff will evaluate the invention (sometimes but not always in a timely manner) and will file a patent application for the government if it seems in the public interest and is practical to do so. If the government obtains a patent, the organization may retain a nonexclusive, royalty-free license and the inventor may receive royalty payments according to a standard formula. If the agency elects not to exercise the government's rights to the invention, these rights may be granted back to the inventor, who may then file for a patent.[46]

As recipients of federal funding, nonprofit organizations and small business firms are subject to the regulations contained in 37 CFR § 401, Rights to Inventions Made by Nonprofit Organizations and Small Business Firms. All applicable organizations need to fulfill certain initial reporting requirements; give the government an irrevocable, royalty-free, nonexclusive license for governmental purposes; and follow certain other related reporting and diligence requirements, including the following:

PATENT APPLICATION and ACKNOWLEDGMENT: At the time the organization or the inventor submits the formal application to the U.S. PTO, a copy also should be sent to the Extramural Invention Reports Office at the NIH, along with the obligatory license. The patent application must include the following statement: "This invention was made with Government support under (identify the grant/contract) awarded by the (cite the awarding agency, for example, the National Institutes of Health). The Government has certain rights in the invention,"

LICENSE: Every patent applicant (individual or institutional) is required to provide the government with a nonexclusive, irrevocable, paid-up license in the invention. A single copy of this license should be sent to the report office within the agency (e.g., the Extramural Invention Reports Office at the NIH), when the patent application is filed.

GRANTEE/CONTRACTOR: The grantee/contractor may retain the entire rights, title, and interest throughout the world to each invention in accordance with the provisions of 35 U.S.C. § 203. With respect to any subject invention to which the grantee/contractor retains title, the federal government shall have a nonexclusive, nontransferable, irrevocable, paid-up license to practice or have practiced for or on behalf of the United States the subject invention throughout the world.

Each federal government agency provides a guide for grants and contracts that can be obtained on its Web sites. In addition, *The NIH Guide for Grants and Contracts* reference 37 CFR § 401.14. Government compliance requirements are explained further in *The AUTM Manual*.[46]

LICENSING

A license is the permit granted by the owner of an intellectual property (be it a government agency or an academic or nonprofit organization) to a person, firm, or corporation to make, use, and/or sell the product or process based on the owner's invention(s) to pursue commercial development. In other words, a license is a contract in which the licensor (owner) grants to a licensee (buyer/ developer) certain rights to a specified property belonging to the licensor.[47] A license is governed by state or federal law, depending on whether the owner of the invention is a government agency or not. The license represents a promise from the owner not to sue the licensee for infringement; it gives permission to do what—for the lack of a license—is unlawful.[48]

An inventor of a patented product or process may assign or sell all rights to that invention. The sale or assignment of a patent is a transfer of property;[49] this transaction prohibits the rest of the world, including the inventor, from making, using, or selling embodiments of the patented invention. However, if the inventor licenses someone to make, use, and sell the patented invention, no one is being prohibited from doing anything; instead, the licensee is simply being given the right to infringe on the patent without being sued.[50] Unlike a patent assignment, which conveys all right, title, and interest to the patent for the full life of the patent, a license may be granted for a period shorter than the remaining term of the patent.[49]

Licensing, then, is a way to transfer to another party the right to use an invention in return for negotiated fees, royalties, and/or equity.[50] The purposes of licensing an invention:

1. to provide a mechanism for transferring the results of an organization's research to the public for the public good; and
2. to generate income for the organization's mission.[50]

In most academic, nonprofit organizations and government agencies, net proceeds (also known as royalties) from licensing income are shared by the inventor and the organization/agency in accordance with the latter's patent policy. Usually, the organization's share is used to finance patent expenses and to support research within the organization. The inventor's share is considered personal income and needs to be listed as such on income tax forms. A technology transfer office can provide a copy of the institute's patent policy on an inventor's distribution of royalty.

There is no standard license since the programs and disciplines within an organization vary, as does the patented technology. The provisions of licenses vary greatly, depending on:

1. the type of invention involved;
2. the commercial investment required to exploit the invention;
3. whether government funding has been involved; and
4. the need for approval from federal/state regulatory agencies.[51]

There are two main types of licenses: exclusive and nonexclusive. An exclusive license grants only one licensee rights to an invention, while a nonexclusive license can site any number of licensees. An exclusive license does not always give the licensee sole rights to the entire invention. Instead, it may grant the rights to a certain field of use for the invention (e.g., diagnostic, therapeutic or device) and/or it may grant rights restricted to a certain market share (e.g., the United States, Europe, or Japan versus worldwide). Thus an invention can be licensed exclusively to a number of licensees but only in a limited field and/or only for a limited time.[49]

Usually, an exclusive license involves a technology that is or will be the end product and thus will require not only FDA approval but also large capital expenditures or additional research and development. A nonexclusive license, on the other hand, usually involves component technologies that are ready for marketing with relatively few additional expenditures. Research reagents or tools and unpatented biologicals are examples of materials that are most often licensed nonexclusively.[49]

Each organization has its own exclusive and nonexclusive license agreements, with some degree of variation in nonexclusive formats. A Commercial Evaluation License allows an interested licensee to "test" patented or patent pending materials, in their own laboratory conditions for a certain period and, for a fee, before deciding on a formal licensing agreement. A Commercial In-House License is usually for a specific use of material(s) as a component, not as a final product. Generally, a period of use is specified, with an annual fee requested for the term. A Biological Material License is for unpatented biological materials and usually entails a one-time fee.

A number of variables are considered during license negotiations: Which potential licensee is most likely to bring the product to market? Does the invention have broad application so that it could be licensed nonexclusively? Or is it a single product that is best licensed exclusively? What is the size of the market for the product? What will be the financial terms of the license? These factors and others affect the final license.[48] Despite variations, however, some provisions are common to most general technology licensing agreements:

1. Definitions: Licensed Products, Licensed Patents, Licensed Trade Secrets, and Know-How
2. Grant clauses: (a) exclusive/nonexclusive; (b) right to make, have made, use, and sell; (c) sublicensing
3. Technological information and assistance
4. Royalties (upfront fee and/or minimum annuals, percentage of net sales, etc.)
5. Improvements made by licensee
6. Duration and termination
7. Miscellaneous provisions (warranties, disclaimers, government regulations)
8. Foregone (abandoned) patents
9. Termination problems

As mentioned above, in an exclusive license agreement, the licensor provides to the licensee an exclusive right to make, use, sell, and practice the invention as well as the right to enforce the patent. In return, the licensee agrees to meet benchmarks and product development milestones; make royalty payments; if appropriate, meet U.S. manufacturing requirements; and, if applicable, sub- or cross-license rights to allow future sponsored research or CRADAs. A licensor can modify or terminate the license agreement because of the licensee's failure to meet benchmarks; execute the commercial development plan; keep the product reasonably available to the public after commercial use commences (if required); and reasonably satisfy unmet health and safety needs. In acquiring a license, especially an exclusive license, the licensee is making a commitment that, if not met, may have legal and financial repercussions.

In order to attract potential licensees, institutions make their inventions known in Non-Confidential Disclosure Statements. These statements reveal enough about an invention to interested licensees without disclosing sufficient information to enable someone to practice the invention.[52] An interested potential licensee usually requests more detailed information. Under this circumstance, a Confidentiality Agreement signed by the institution and by the potential licensee is necessary to protect the inventor's, the institution's, and the potential licensee's interests. Licensing can take place any time after the disclosure of an invention; i.e., after either publication or filing of a provisional/patent application. The inventor is usually a good resource in identifying potential licensees.

However, all requests from companies for information about an invention should be directed to the technology transfer office.

ROYALTY AND EQUITY

Royalty is a payment the licensor collects from a licensee as part of a license.[49] A license royalty can involve five types of payments collected at various times:

1. an execution fee usually collected 60 days after the license agreement has been signed by both parties;
2. a minimum annual fee collected each year on the anniversary of licensing for the term of the license;
3. benchmark royalties (milestone payments) based on the attainment of certain goals in product development of the licensed technology (e.g., filing of an IND application, first commercial sale, or patent issues);
4. patent costs through which the licensee reimburses the licensor for domestic and/or foreign patent costs; and
5. earned royalties, a fixed percentage of annual net sales.

Not all licenses entail all five types of royalties. Each license is different and technology dependent. Generally, an exclusive license entails all (but sometimes not, patent costs), while a nonexclusive license usually does not cover patent costs and may or may not include minimums.

A royalty may be determined in a totally arbitrary manner. Many consider determining royalties a black box. There is no right or wrong royalty rate nor is there a magic formula, although some licensing personnel may have their own calculations for obtaining an estimated royalty. Royalties usually are determined in negotiations between a willing seller and a willing buyer, with the free market system determining a royalty rate. However, use of a well-established approach based on profit sharing and/or cost savings allows the licensor to arrive at a royalty range both parties consider reasonable. Agreement on the amount and the method of payment can require substantive negotiations.[53]

An increasing number of academic and nonprofit institutions are embarking on arrangements involving equity as either a partial or a total source of royalty payments. In fact, according to a 1993 AUTM Public Benefits Survey, 995 companies were formed by U.S. universities and nonprofit institutions between 1980 and 1993. That figure is probably higher now.

Equity is associated with start-up or early-state companies, vehicles of commercialization that are commonly cash poor.[54] Equity may be provided as part of an upfront licensing fee, as partial or total replacement of royalty fees, and as an incentive for inventor/faculty participation in the development of the research project and/or as consulting fees.[54]

Academic and nonprofit institutions become involved with creating a company for a number of reasons. For example, the formation of a company can increase the probability of successful commercialization of technology, promote and support local economic development, satisfy career goals of scientists/inventors, and carry prestige. Moreover, equity allows the institutions to share in creating value. However, the creation of companies by institutions also raises concern about issues such as the appropriate consideration for the value of the technology and the other contributions of the institution, the potential conflict of interest of the scientific founders and the institution, and academic freedom or rights to future research and improvements.

Taking equity:[54]

1. enables a new company to conserve cash reserves;
2. captures the value of an entire company, not just a single technology;
3. recognizes the larger contribution of the institution to the value of the company;
4. may enable the institution to realize the value from a license prior to product sales.

Unlike other licensing arrangements, in acquiring equity, the institution must guard against conflicts of interest. The PHS offers guidelines for the management of such conflicts. A major concern for institutions is monitoring the equity inventor/scientist for conflicts of interest versus conflicts of commitment. Institutions have developed policies on accepting equity and balancing a company's interests with academic freedom. In some institutions, committees are established to review equity cases for conflicts on an annual basis.

The current climate at academic and nonprofit institutions is fertile for creating companies. With grant funding becoming more difficult to obtain, alternative sources of funding are needed and closer relationships are thus being established between research institutions and industry. In addition, the PHS/NIH conflict of interest rules and institutional policies are allowing for more flexibility.

In short, the licensing to start-up companies is an increasingly accepted mechanism of commercialization for academic and nonprofit institutions.[55,56] Accepting equity as part of the consideration for licensing is common and is frequently necessary in these arrangements. The appropriateness of equity participation should be evaluated for each case; points to be considered include potential conflicts of interest and commitment, adequate management, sources of capital, and nature of the technology.[54] Information on an institution's equity policy may be obtained from its technology transfer office.

FEDERAL TECHNOLOGY PROGRAMS

The federal government provides funding programs for U.S.-developed technologies. Of these programs, two in particular—the Small Business Innovation Research (SBIR) and the Small Business Technology Transfer Research (STTR)—have been successful in promoting research between academic/nonprofit institutions and small U.S. businesses.

The SBIR program (see Appendix D) is a highly competitive three-year award system providing qualified small businesses with the opportunities to propose innovative ideas that meet the specific R&D needs of the federal government. Each year has a certain amount of funding and certain requirements. The SBIR program is designed to assist participating federal agencies in meeting their respective R&D needs and to provide qualified small businesses with opportunities to compete for a greater share of federal R&D awards. Each participating federal agency designates topics for SBIR proposals, releases at least one SBIR solicitation annually, and receives and evaluates SBIR proposals at designated times throughout the year. SBIR awards are based on technical and scientific merit, cost-effectiveness, and the agency's needs and requirements.

The principal investigator (PI) of an SBIR must be primarily employed by the small business applying for the award. In Phase I of the award, 67% of the entire research plan must be carried out by the PI of the small business, with not more than 33% undertaken by an institution as subcontractors or consultants. In Phase II of the award, at least 50% of the research plan generally must be carried out by the small business, with the remaining 50% performed by academic/nonprofit institutions. Both Phase I and Phase II must be performed entirely in the United States.[57]

The STTR program (see Appendix E), a federal funding program for R&D established in October 1993, promotes joint research between academic/nonprofit research institutes, including some federal laboratories, and U.S. small businesses. This three-year pilot program facilitates joint research and development. It is being considered by Congress for renewal. However, each project must be submitted by the U.S. small business that will conduct at least 40% of the research. The academic/nonprofit partner is a subcontractor under the grant and must perform 30% of the research, with the remaining 30% negotiable. The STTR program's Phase I award is for 12 months (as opposed to 6 months for the SBIR program).[58]

The five federal agencies required to participate are those with extramural R&D budgets in excess of $1 billion per year: the Department of Defense, the National Institutes of Health, the Department of Energy, the National Aeronautics and Space Administration, and the National Science Foundation. The STTR and SBIR programs are managed together in the participating agencies. An application, a list of proposals, and detailed information on these two programs can be obtained by the U.S. Small Business Administration and from the federal agency of interest.

TECHNOLOGY TRANSFER OFFICES

How does technology transfer work? At each institution, technology transfer is implemented differently, depending on the mission of the organization. However, an established office or at least one person is responsible for assisting faculty members, scientists, and research staff members with the mechanisms of technology transfer. It is often a pleasant surprise to find how knowledgeable and helpful this office can be.

Certain functions are standard in any technology transfer office, no matter how large or how small. The major services that such an office provides are:

Disclosure of inventions
Record keeping and management
Evaluation and marketing
Patent prosecution
Negotiation and drafting of license agreements
Management of active licenses

Technology transfer offices have many different customers whose objectives sometimes conflict with one another. For instance, customers may include:

1. faculty-inventors, who often have expectations regarding research opportunities, income, public utilization, and fame;
2. the private sector, which expects to secure commercially viable technology at a fair price;
3. the university administration, which expects the office to be self-supporting and wants to prevent conflicts of interest;
4. the governing board, director, and/or president of the institution, who may need assurance that the institution's name and reputation are protected in its industrial relationships and that the missions of the institution (including, education, research, and service) are not compromised by business interests;
5. the taxpayers, who expect the office to manage state and federal resources and research in an effective and nondiscriminatory manner; and
6. the sponsoring government agency, which insists on compliance with provisions of the Bayh-Dole Act.

One point needs to be mentioned: without research, there is no technology transfer and thus no benefit to the public. Interested researchers, faculty and/or staff, and technology transfer personnel who are not afraid of exploring the possibilities of technology transfer, can work together to create a useful and valuable technology transfer office. Given all that is involved in running a technology transfer office, the process of technology transfer takes time and perseverance. Nothing happens overnight! Patience combined with the right amount of intensity is required from both you and your technology transfer colleague.

All aspects of research and the people involved in those fields represent the key to transferring results into cures, while the technology transfer office in your institution provides the means to unlock the resources. Go and visit your colleague in your technology transfer office to find out what you always wanted to know about technology transfer but were afraid to ask.

CONCLUSION

Scientists in government, university, and nonprofit laboratories are an abundant source of today's innovative ideas that become tomorrow's cures. Understanding the laboratory-to-market mechanism is the first step in the transfer of inventions to the commercial world for the public good.

The technology transfer business has changed considerably in the last five years. The field has grown larger and more competitive as increasing numbers of research institutions establish programs to patent their inventions and license them to industry. Researchers are not alone in the technology transfer process. The technology transfer office provides the expertise and guidance necessary for negotiating "the technology transfer maze," thus unlocking resources and harvesting their benefits.

Appendix A

Technology Transfer Terminology

1. Application

An application of specified form and content is filed with the U.S. PTO and requests that the government grant a property right (patent) for a specified invention.

2. CIP

A Continuation-In-Part (CIP) patent application is filed when, at some point after filing an initial application, the inventor discovers an improvement or a further use for, or additional data confirming a debatable aspect of the invention.

3. Claim

A claim is the legal description of an invention in a patent application or an issued patent. The claim defines the property with regard to which the patent grants the owner the right to exclude all others.

4. Copyright

Copyright is the exclusive property right granted by the government to authors, artists, and others to prevent third parties from copying original literary, artistic, or other work. Like a patent, copyright is granted for a specific term after which it expires and the work enters the public domain.

5. Divisional

When a utility patent application is filed and the U.S. PTO determines that more than one invention is represented by that application, the PTO can require the applicant to choose which invention to prosecute first. A divisional patent application can then be filed for each of the other inventions. A divisional application is similar to a regular utility patent application except that its filing date is the same as that of the original utility patent application.

6. Filing Date

The filing date is the date on which a patent application is received at the U.S. PTO. This date establishes what is prior art to this application.

7. Infringement	Infringement is the unauthorized use of a property right owned by another. Infringement results only from activity by one party within the property rights of another, such as the claims of a valid patent. A patent or other publication alone cannot infringe a patent; it is the act of making, using, or selling of a patented invention without consent of owner that is required.
8. Intellectual Property	Intellectual property includes patents, trade secrets, trademarks, trade names, and copyrights.
9. Invention	An invention is a new technical development that may or may not be patentable.
10. License	A license is a permit granted by a governmental agency or an academic, nonprofit, or for-profit organization to a person, firm, or corporation to make, use, and sell an invention without the transfer of ownership of the invention in order to pursue commercial development of a product that is patented, has a patent pending, or is nonpatented. The license may be exclusive or nonexclusive (see text). If the requested material is patented or has a patent pending, a Commercial Evaluation License, a Commercial Research License or a Patent License Agreement will be granted. If the requested material is nonpatented, then a Biological Material License Agreement is considered.
11. Patent	A patent is a contract created by law. In return for the disclosure of an invention, the government grants the owner property rights that exclude all others from using the invention, provided that the government finds the invention patentable under the established law. The government grants the property rights for a specified number of years (17 from issuance for plants; 14 from issuance for design; 20 from filing date for utility), after which the patent expires and the invention enters the public domain.
12. Patentability	An invention may or may not be patentable under the patent laws governing the patent ap-

plication or patent in question. To be patentable, an invention must be novel, unobvious, useful, and enabling.

13. Prior Art

The term prior art refers to the information in a particular field that is published or generally available before a specified date, usually the date an application is filed with the PTO. Whether a prior patent or publication is prior art is determined by the patent law under which the patent application or the issued patent is being evaluated. Prior art may be in the public domain, or it may be a property right of the owner of an unexpired patent.

14. Provisional Application

An inventor may file a provisional application on an invention. This informal document, which does not require claims or formal filing papers, preserves rights in the invention for up to one year, at the end of which a standard patent application must be filed if the inventor wishes to obtain a patent. During this year, no formal procedures are undertaken by the PTO with regard to the provisional application. The provisional filing date may also be used as a priority date for foreign filing. The cost of filing a provisional application at the PTO is low compared with that of the standard patent application.

15. Public Domain

Information in the public domain is generally available information for which no one has a property right.

16. Serial Number

A seven-digit serial number is assigned to a patent application once it is received or "docked" at the PTO. This number identifies the patent application throughout its prosecution and is abbreviated SN.

17. Trademark

A trademark is the identifying mark, word, logo, or symbol used by an institution in commerce to identify or distinguish its goods and services from all others.

18. Trade name

A trade name is the company or firm name under which a business is conducted.

19. Trade Secret

Trade secret is intellectual property protected by keeping the information secret, usually giving a manufacturer an advantage over competitors.

20. U.S. PTO   The United States Patent and Trademark Office (Crystal City, VA) is the government agency under the Department of Commerce that is responsible for all activities related to the patent process, including filing and reviewing patent applications as well as issuing and maintaining patents.

Additional term definitions can be obtained in the *Technology Transfer Guide-1994 Edition*[3] Exhibit A, p. 1, as well as *The AUTM Technology Transfer Practice Manual*, Volume II, Part X: Appendix A, p. 1.

Appendix B

Material Transfer Agreement
(Provider)

The "Institution," located at _____, agrees to provide certain materials and information ("Materials") for the purpose of scientific collaboration to

_____
(name of principal investigator)

_____
(institutional affiliation and address)

("Recipient") for the purpose of

_____
The Materials provided by the Institution to the Recipient under this Agreement are:

_____

The Materials are provided under the following conditions, which are agreed to by the Recipient:

1. Ownership. All Materials, including progeny and derivatives thereof, are and shall remain the property of the Institution (subject only to whatever rights the United States government may have to the Materials). Nothing contained within this Agreement restricts the Institution's rights to such Materials, including the Institution's rights to use or distribute the Materials to other commercial or non-commercial entities.

2. Use. The Recipient agrees that the Materials: (i) shall be used only by the Recipient and only for the research purposes described above; (ii) shall not be used in human subjects; and (iii) shall not be used, directly or indirectly, for commercial purposes. If the Recipient creates material(s) derived from the Materials provided which becomes or is used in the development of a commercial product, then the appropriate license agreement will be acquired from the Institution.

3. Distribution and Control. The Materials represent a significant investment on the part of the Institution and are proprietary to the Institution. The Recipient shall maintain the confidentiality of Materials and agrees not to transfer or disclose the Materials to any third party without the prior written permission of the Institution. Upon the Institution's request, the Recipient shall return all Materials to the Institution, retaining no part thereof. In addition, the Recipient

shall obtain acceptance of the terms of this Agreement from all persons who have access to the Materials.

4. Patent. If in any way the use of the Materials results in any inventions, improvements and/or ideas, whether or not patentable, the Recipient shall promptly disclose and assign to the Institution said inventions, improvements, and/or ideas. If the Institution should elect to pursue patent protection for said inventions, the Recipient shall cooperate with the Institution in furthering the filing and prosecution of patent applications for such inventions and shall ensure that its employees and agents cooperate with the Institution in this regard.

5. Reporting. The Recipient shall supply to the Institution a draft of any publication contemplated for submission or any proposed public disclosure resulting from the use of the Materials no later than sixty (60) days prior to its submission, and no later than ninety (90) days prior to its public disclosure. The Institution shall treat such draft confidential until publication or disclosure but may include information from such draft in patent applications, as appropriate.

6. Warranty. The Recipient accepts the Materials with the knowledge that they are experimental and agrees to comply with all laws and regulations for the handling and use thereof. Because the Materials are experimental, IT IS UNDERSTOOD THAT THEY ARE BEING PROVIDED WITHOUT WARRANTIES, EXPRESS OR IMPLIED, INCLUDING ANY WARRANTY OF MERCHANTABILITY OR FITNESS FOR A PARTICULAR PURPOSE OR WARRANTY AGAINST INFRINGEMENT.

7. Liability. The Recipient hereby waives any claim against the Institution and further agrees to indemnify, defend, and hold the Institution harmless from and against any and all claims, suits, losses, damages, liabilities, and expenses, including reasonable attorneys' fees, which may be alleged to arise out of or in connection with the Recipient's receipt, use, disposition, handling, or storage of the Materials.

8. Export. The Recipient agrees that it will not knowingly export or re-export any technical data or Materials furnished or resulting from the provision of Materials, without first obtaining permission to do so from the Institution and from the U.S. Department of Commerce, the U.S. Food and Drug Administration, and/or other appropriate governmental agencies as may be required by law. The Recipient further agrees that it will at all times comply with all applicable federal, state and/or local laws or regulations, including, but not limited to, the National Institutes of Health Guidelines.

The Institution and the Recipient have caused this Agreement to be executed by their respective duly authorized officers. This Agreement shall be effective as of the date last set forth below.

INSTITUTION   **RECIPIENT**

By: _____   By: _____
Name: _____   Name: _____
Title: _____   Title: _____
Date: _____   Date: _____

RECIPIENT'S INSTITUTIONAL APPROVAL BY:

By: _____
Name: _____
Title: _____
Date: _____

* Adapted from The National Institute of Health's and The Children's Hospital of Philadelphia's Material Transfer Agreement.

Materials Receipt Agreement

The "Institution," located at _____,
agrees to receive certain materials and information ("MATERIALS") from:

_____ ("PROVIDER").

The MATERIALS provided to the Institution by the PROVIDER under this
Agreement are:

_____

The MATERIALS are provided under the following conditions:

1. All MATERIALS are and shall remain the property of the PROVIDER.

2. Upon the PROVIDER's request, the Institution shall return all MATERI-
ALS to the PROVIDER.

3. The Institution shall not distribute the MATERIALS to anyone not having a
confidential relationship with the Institution, without the written consent of the
PROVIDER.

4. The Institution shall supply to the PROVIDER a copy of any publication
resulting from the use of the MATERIALS. If provided in pre-publication or
manuscript form, all such communications shall be maintained by the PROVIDER
as confidential until publication or disclosure.

5. The Institution accepts the MATERIALS with the knowledge that they are
experimental and agrees to comply with all laws and regulations for the handling and use thereof. Because the MATERIALS are experimental, it is understood that they are being provided without warranties, express or implied,
including any warranty of merchantability or fitness for a particular purpose.

6. The Institution shall not permit the use of the MATERIALS, either directly
or indirectly, in humans without obtaining appropriate IRB and/or regulatory
clearances.

7. Any inventions, improvements or ideas, whether or not patentable, stem-
ming from or in any way related to the Institution's use of the MATERIALS
shall belong to the Institution. The Institution has a proprietary interest in the
field of _____ to be used with the MATERIALS and recognizes the
PROVIDER's own interest with the MATERIALS. In the event that the Institution shall file a U.S. patent application relating to the use of such MATERIALS,
the Institution shall provide a copy of such patent application to the PROVIDER
on a strictly confidential basis. The PROVIDER may, at its option, request to
negotiate an exclusive or nonexclusive license to such patent application and
any patent issuing thereon. The Institution shall then enter good-faith negotiations with the PROVIDER concerning such license, provided such patent application is not subject to a previous obligation to license, is not previously licensed,

and is not otherwise encumbered in a manner which precludes the grant of such a license to the PROVIDER.

Intending to be legally bound, the Institution and the PROVIDER agree to the foregoing terms.

INSTITUTION   **PROVIDER**

By: _____   By: _____
Name: _____   Name: _____
Title: _____   Title: _____
Date: _____   Date: _____

RECIPIENT's SCIENTISTS:

Name: _____
Date: _____

* Adapted from the National Institutes of Health's and the Children's Hospital of Philadelphia's Material Transfer Agreement.

Appendix C

Public Law 104-113, The National Technology Transfer and Advancement Act of 1995

Public Law 104-113 (P.L. 104-113) is a bill to amend the Stevenson Wydler Technology Innovation Act of 1980 (P.L. 96480) and the Federal Technology Transfer Act of 1986 (P.L. 99502). This law provides for the following key issues (8):

- It permits government laboratories to use funds received from a collaborator in a CRADA to hire Full-Time Equivalent (FTE) ceiling exempt personnel to assist in the CRADA.
- It permits the federal government to reassign title to patent applications back to the inventor if the government decides to discontinue patent prosecution and otherwise not pursue commercialization.
- It permits government laboratories to use license royalty revenue for mission related research in the laboratory, for technology transfer related administrative and legal costs, for promotion of scientific exchange among agency laboratories, and for education and training.
- It provides the inventive entity with the first $2000 of royalties and at least 15% of the royalties per year accrued for inventions made by the inventor. (According to NIH policy, the laboratory and/or inventors receive 25% of the first $50,000, 20% of the second $50,000, and 15% of all royalty income thereafter.
- It increases the individual maximum royalty award to $150,000 per year.
- It provides a royalty free right to use sole-collaborator CRADA subject inventions for legitimate government needs.
- It permits federal government laboratories to require sublicensing of CRADA inventions to others in exceptional circumstances for compelling public health, safety, or regulatory needs while providing administrative appeal and judicial review of the agencies' requirement in such rare circumstances.
- It provides financial support of over $200,000 annually to the Federal Laboratory Consortium (FLC).

Note: H.R. 2544, The Technology Transfer Commercialization Act of 1997 was recently introduced as a bill that would promote technology transfer by facilitating licenses for federally owned inventions. This bill would: (1) amend the Bayh-Dole Act by removing the legal obstacles to effectively licensed federally owned inventions created in government-owned, government operated (GOGO) laboratories, which would parallel authorities currently in place for licensing university or university-operated federal laboratory inventions; (2) amend The Stevenson-Wydler Act to allow federal laboratories to include already existing patented inventions in Cooperative Research & Development Agreements (CRADAs); and (3) remove the language requiring public notification procedures on a government invention for exclusive licensing in The Federal Register. (Adapted from Hon. Constance A. Morella remarks of September 25, 1997 in The Congressional Record Page: E1851.)

Appendix D

Small-Business Innovation
Research (SBIR) Program

Purpose: Commercialization of research project.

Three Phases: Phase I: The objective is to determine the scientific and technical merit, feasibility, and potential for commercializa-
tion of the proposed project and the quality of the
performance of the small-business concern before consideration of further federal support (in Phase II). Generally, no more than 1/3 of the project may be
conducted by consultants and contracts in Phase I.

Funding: *Up to $100,000* for direct costs, indirect costs,
and negotiated fixed fees for a period normally *not to
exceed six months.*

Phase II: The objective is to continue the research efforts from
Phase I. Funding is based on the results achieved in
Phase I and the scientific and technical merit and commercial potential of the Phase II proposal. Generally,
no more than 1/2 of the project may be conducted by
consultants and contracts in Phase II.

Funding: *Up to $750,000* for direct costs, indirect costs,
and negotiated fixed fees for a period normally *not to
exceed 2 years;* i.e., in general, a 2-year project may not
cost more than $750,000.

A Phase I award must have been received in order to
apply for a Phase II award.

Phase III: The objective is to pursue, *with non-SBIR funds,* the
commercialization of the results of the research project
funded in Phases I and II.

Eligibility: For-profit small-business concern (sole proprietorship,
partnership, corporation, joint venture, etc.) with no
more than 500 employees.

Economically and socially disadvantaged small businesses and woman-owned small businesses are encouraged to participate but are given no preferential
treatment.

Principal Investigator: Designated by small business to be responsible for scientific and technical direction of research plan. This

person must be primarily employed (> 50%) by small business at time of award and during conduct of project.

Dates: Grant application due on April 1, August 1, and December 1.

Appendix E

Small Business Technology Transfer Research (STTR) Program

Purpose:	To facilitate cooperative research and development (R&D) with potential for commercialization between small-business concern and U.S. research institution.
Three Phases:	Same as for SBIR program.
Cooperative R&D:	In both Phase I and Phase II, at least 40% of the project (both Phases I and II) must be performed by the small business concern and at least 30% of the project must be performed by the research institution (college or university, other nonprofit research organization; or federal R&D center but not laboratories staffed by federal employees).

Funding: Phase I: Up to $100,000 for direct costs, indirect costs, and negotiated fixed fees for a period normally not to exceed 1 year.

Phase II: Up to $500,000 for direct costs, indirect costs, and negotiated fixed fees for a period normally not to exceed 2 years; i.e., in general, a 2-year project may not cost more than $500,000.

A Phase I award must have been received in order to apply for a Phase II award.

Eligibility:	For applicant organization: same as for SBIR program. The applicant small-business concern will be the recipient and will execute a subcontract with the research institution for that institution's performance under the STTR award.
Principal Investigator:	Same as for SBIR program, EXCEPT that the PI may be primarily employed by an entity *other* than the small-business concern, including the research institution.
Dates:	Grant application due on April 1, August 1, and December 1.

REFERENCES

1. *Technology Transfer Guide-1994 Edition*, Grants Administration News Co., Plano, TX, Exhibit D, pp. 1–2.
2. Adapted from Elaine V. Jones AAI Symposium on Partners in Science, June 12, 1996.
3. *Technology Transfer Guide-1994 Edition*, Grants Administration News Co., Plano, TX, Exhibit D, pp. 1–2.
4. 35 U.S.C. § 200 et seq.
5. 15 U.S.C. § 3701 et seq.
6. *Technology Transfer Guide-1994 Edition*, Grants Administration News Co., Plano, TX, Exhibit D, p. 2.
7. The Federal Technology Transfer Act of 1986 (FTTA), Public Law 99-502, 15 U.S.C. § 3710.
8. Public Law 95-521, October 26, 1978, 92 Stat 1824, et seq.
9. M. Witt and L. Gostin, *JAMA*, 27(7):547, 1994.
10. *AUTM Technology Transfer Practice Manual*, Volume II. Chapter 2: Uniform Biological Material Transfer Agreement(s) Joyce Brinton, The Association of University Technology Managers, Inc., February 1996, p. 1.
11. Ibid., p. 2.
12. Adapted from C. Dietzel talk on "University-Industry Relationships: Partners in Science."
13. *AUTM Technology Transfer Practice Manual*, Volume II, Part IX, Chapter 8: Cooperative Research and Development Agreements (CRADAs) Daniel Massing, The Association of University Technology Managers, Inc., February 1994, pp. 1–50.
14. United States Public Health Service, *Technology Transfer Manual*, Chapter No. 402, "National Institutes of Health's Cooperative Research and Development Agreement Procedure," 1997, p. 2.
15. Technology Transfer-Federal Agencies: Technology Patent Licensing Activities, GAO/RCED-91-80, April 1991.
16. *AUTM Technology Transfer Practice Manual*, Volume II, Part IX, Chapter 8: Cooperative Research and Development Agreements (CRADAs) Daniel Massing, The Association of University Technology Managers, Inc., February 1994, p. 3.
17. *Ratner & Prestia Insight*, 5(1):1–2, 1994.
18. Personal Communication with Dr. Kathleen Sybert, Office of Technology Development, National Cancer Institute, National Institutes of Health, June 20, 1996.
19. *Technology Transfer Guide-1994 Edition*, Grants Administration News Co., Plano, TX, Exhibit B, pp. 1–2.
20. *AUTM Technology Transfer Practice Manual*, Volume I, Part I, Chapter 2: A Patent—What Is It and How Do You Get It? Charles Van Horn, The Association of University Technology Managers, Inc., February 1996, pp. 1–7.
21. Ibid., pp. 6–7.
22. Ibid., pp. 1–3.
23. 35 U.S.C. § 154.

24. 35 U.S.C. § 102.
25. 35 U.S.C. § 101.
26. 35 U.S.C. § 103.
27. 35 U.S.C. § 112.
28. 37 CFR § 1.51.
29. 37 CFR § 1.71.
30. *Manual of Patent Examining Procedure* (MPEP) 608.01(p), 1995.
31. 37 CFR § 1.75.
32. *Intellectual Property Today*, 4(1):16–17, 1995.
33. *AUTM Technology Transfer Practice Manual*, Volume 1, Part, I, Chapter 2: A Patent—What Is It and How Do You Get It? Charles Van Horn, The Association of University Technology Managers, Inc., February 1996, pp. 16–28.
34. 37 CFR § 1.141–1.142 and 1.146.
35. 37 CFR § 1.20.
36. 35 U.S.C. § 111(b).
37. *AUTM Technology Transfer Practice Manual*, Volume I, Part I, Chapter 2: A Patent—What Is It and How Do You Get It? Charles Van Horn, The Association of University Technology Managers, Inc., February 1996, pp. 8–9.
38. Lecture and notes from Dale Hoscheit on "Biotechnology Inventions—U.S. Patents," September 15, 1994.
39. *AUTM Technology Transfer Practice Manual*, Volume I, Part IV, Chapter 2.1, Patent Filing in the United States, Kathleen R. Terry, The Association of University Technology Managers, Inc., February 1996, pp. 1–28.
40. Ibid. Chapter 2.2: Guide to Foreign Patent Protection, Steven W. Lundberg, Warren D. Woessner and Ann S. Viksnins, pp. 1–11.
41. 35 U.S.C. § 102(b).
42. *AUTM Technology Transfer Practice Manual*, Volume I, Part I, Chapter 2: A Patent—What Is It and How Do You Get It? Charles Van Horn, The Association of University Technology Managers, Inc., February 1996, p. 14.
43. P. Hider, *Journal of the Association of University Technology Managers*, VI, 49–82, 1994.
44. *AUTM Technology Transfer Practice Manual*, Volume I, Part V, Chapter 2: Documentation of Inventions W. Mark Crowell, The Association of University Technology Managers, Inc., February 1994, pp. 1–4.
45. 37 CFR § 401.
46. *AUTM Technology Transfer Practice Manual*, Volume I, Part I, Chapter 1: Pertinent Laws: Compliance Requirements Jane Youngers, The Association of University Technology Managers, Inc., February 1994, pp. 1–48.
47. *Ratner & Prestia Insight*, 7(2):3, 1996.
48. *Technology Transfer Guide-1994 Edition*, Grants Administration News Co., Plano, TX, Exhibit B, p. 5.
49. Lecture and notes from W.A. Biggart on "Licensing and Technology Transfer Agreements in Biotechnology," November 3, 1994.
50. *The University of California, San Francisco Policy Guide:* 100-25, -26, and -28, Management of Intellectual Property: Patents, Licensing, Transfer of Biological Materials, and Copyright Practices. Regents of the University of California, October 26, 1995.

51. *Technology Transfer Guide-1994 Edition.* Grants Administration News Co., Plano, TX, Exhibit B, p. 6.

52. *The University of California, San Francisco Policy Guide:* 100-25, -26, and -28, Management of Intellectual Property: Patents, Licensing, Transfer of Biological Materials, and Copyright Policies. Regents of the University of California, October 26, 1995.

53. *AUTM Technology Transfer Practice Manual,* Volume II, Part VII, Chapter 3: Royalties, Valuation, Financial Considerations. Marcia Rorke, Edmund Astolfi, and Bernard Friedlander. The Association of University Technology Managers, Inc., February 1994, pp. 1–8.

54. *AUTM Technology Transfer Practice Manual,* Volume III, Part IV, Chapter 2: Equity and Start-Ups Rita C. Manak, The Association of University Technology Managers, Inc., February 1995, pp. 1–11.

55. *AUTM Technology Transfer Practice Manual,* Volume II, Part VII, Chapter 1: Considerations When Starting a Company to Commercialize a University-Developed Technology Michael Moore, A.R. Potami, James Severson, and Anthony Strauss. The Association of University Technology Managers, Inc., February 1995, pp. 1–16.

56. *AUTM Technology Transfer Practice Manual,* Volume III, Part IV, Chapter 3: Formation of a Business Incubator Daniel Massing and Edward Zablocki, The Association of University Technology Managers, Inc., February 1995, pp. 1–19.

57. Department of Health and Human Services *Omnibus Solicitation of the Public Health Service for Small Business Innovation Research (SBIR) Grant Applications,* PHS 95-3, pp. 1–16.

58. Department of Health and Human Services *Omnibus Solicitation of the Public Health Service for Small Business Technology Transfer Research (STTR) Grant Applications,* PHS 95-4, pp. 1–15.

8 Development of a Transmission-Blocking Vaccine: Malaria, Mosquitoes, and Medicine

David C. Kaslow

BRIEF HISTORICAL BACKGROUND

The genesis of a malaria transmission-blocking vaccine and the original discovery of the etiology of the disease malaria are one and the same. In Algeria on November 5, 1880, Dr. Laveran peered through a microscope at the drop of blood recently collected from a 24-year-old French soldier with tertian fevers.[1,2] By withdrawing the blood and thus allowing the temperature to drop to ambient and the carbon dioxide to rise, Laveran had inadvertently mimicked the major trigger that the parasite uses inside the midgut of the female mosquito. The "fine, transparent filaments that moved very actively and beyond question were alive"[2] that Laveran described are the product of exflagellation, now known to be the process by which a male gametocyte emerges from within a circulating red blood cell to form eight male gametes in the mosquito midgut. That malaria parasites were transmitted by mosquitoes was discovered more than a decade later by two scientists working independently, Ronald Ross who was studying the avian malaria parasite *Plasmodium relictum* in India[3] and W.G. MacCallum who was studying avian malaria parasites in the United States and Canada.[4] But it was not until the late 1950s when Huff and his colleagues, working in Bethesda on the avian malaria parasite *Plasmodium gallinaceum*, first demonstrated that vaccination with infected blood could induce transmission-blocking immunity—antibodies that when taken up along with an infectious blood meal by the mosquito could reduce or completely block the infectivity of the parasite to its vector, in this case *Aedes aegypti*.

IDENTIFICATION OF TRANSMISSION-BLOCKING TARGET ANTIGENS

In the mid-1970s, Gwadz[5] and subsequently Carter and Chen[6] showed that vaccination with purified sexual-stage parasites (emerged chicken malaria gametes and zygotes) were sufficient to elicit transmission-blocking antibodies. Funding from IMMAL/TDR/WHO and the advent of three technologies in the 1980s—production of monoclonal antibodies, *in vitro* culture of parasites, and development of membrane-feeding mosquitoes—permitted the first identification of a series of surface antigens of sexual-stage *Plasmodium falciparum* parasites that were targets of transmission-blocking antibodies. In 1985, Ponnudurai and his colleagues in Nijmegen, Netherlands, combined these technologies to identify a target antigen, later called Pfs25, that was cysteine-rich and anchored to the surface of zygotes and ookinetes.[7]

ISOLATION OF THE GENE ENCODING Pfs25

Attempts to clone the gene encoding Pfs25 by immunoscreening of prokaryotic expression libraries had failed, mainly because the highly reducing environment within *Escherichia coli* is not conducive to the recreation of disulfide bonds and, as I will describe below, the recombinant protein product is toxic. In 1986, the most obvious alternative was to purify enough protein to determine an amino acid sequence from which to synthesize degenerate synthetic oligonucleotides. The oligonucleotides were then to be used to screen genomic or cDNA libraries. I took the same approach in two malaria models simultaneously— immunoaffinity chromatography followed by SDS-PAGE to purify approximately 10–50 micrograms of 25-kDa protein from 50–100 infected chickens (*P. gallinaceum* zygotes) and from three months' worth of *in vitro* culture (*P. falciparum* zygotes). The first protein microsequenced was a contaminant, the light chain of immunoglobulin, which also migrated at approximately 25-kDa. To distinguish between the immunoglobulin used in purification and parasite-produced protein, the bulk parasite extracts were spiked with metabolically ^{35}S-labeled protein. From another 50–100 chickens and three months of *in vitro* culture, highly purified parasite-produced protein was eluted from gel slices and microsequenced by automated Edman degradation—unfortunately, the amino-terminus of the mature protein was blocked. Another round of purification to produce tryptic fragments for sequencing internal peptides revealed that the protein was resistant to trypsin unless it had been treated with a denaturing agent and subsequently reduced and alkylated. Finally, three tryptic fragments had been generated from reduced and alkylated protein, and the first round of synthetic degenerate oligonulceotides were used for screening genomic libraries. After screening hundreds of thousands of bacteriophage plaques and plasmid colonies and performing a number of Southern and Northern blots, it was clear that the sequence was not present in the malaria parasite genome. Reanalysis of the original microsequence data revealed that an error had been made in reading one of the amino acid residues. The next round of synthetic oligonucleotides was found to be too degenerate for screening genomic libraries, but it did hybridize specifically to an abundant transcript by Northern blot analysis. After the degeneracy of the oligonucleotide was reduced, the gene was isolated and its sequence determined.[8]

RECOMBINANT PROTEIN EXPRESSION

A search of protein databases with the deduced amino acid sequence from Pfs25 revealed homology to a series of diverse proteins that all contained epidermal growth factor (EGF)-like domains. Between the amino terminal secretory signal sequence and the short hydrophobic region at its carboxy-terminus, Pfs25 had four such domains (the first of which is truncated). EGF-like domains consist of six cysteine residues that form three disulfide bonds. All transmission-blocking monoclonal antibodies to Pfs25 available at the time in my laboratory recognized disulfide bond-dependent epitopes. Recreating the disulfide bonds in a recombinant protein that had 20 cysteines seemed unlikely, but nevertheless I was hopeful that perhaps a linear epitope existed, antibodies to which might mediate transmission-blocking activity. As prokaryotic expression seemed the simplest means of producing recombinant protein, initial attempts were made to express full-length Pfs25 as a nonfused protein and as a *TrpE* fusion protein in *E. coli.* After months of failing to create a recombinant plasmid that encoded full-length Pfs25, experiments in which a series of truncated Pfs25 sequence were used revealed that the signal sequence combined with the carboxy-terminal hydrophodic region was highly toxic to *E. coli* even in tightly regulated expression systems such as *TrpE.* The nonfused truncated protein was expressed at extremely low levels; however, milligram amounts of truncated Pfs25 TrpE fusion protein, as inclusion bodies, were made. Although transmission-blocking mAbs did not recognize the fusion protein, antisera made in mice and rabbits vaccinated with the fusion protein recognized native parasite-produced Pfs25 by live indirect immunofluorescence and immunoblots of nonreduced parasite extracts. None of the sera generated had any transmission-blocking activity.[9]

In an attempt to make recombinant antigen recognized by transmission-blocking antibodies, a recombinant WR strain of vaccinia virus, vSIDK, was made that encoded full-length Pfs25. A series of transmission-blocking mAbs recognized the surface of live vSIDK-infected mammalian cells expressing Pfs25. Despite eliciting antibodies to Pfs25, mice vaccinated once with vSIDK did not elicit transmission-blocking antibodies; however, mice vaccinated three times with vSIDK elicited potent transmission-blocking antibodies.[10] Unfortunately, the virulent WR strain is not suitable for use in immunocompromised humans, and because some of the target population for malaria transmission-blocking vaccines has a high prevalence of HIV positivity, attenuated live organisms or a subunit vaccine approach had to be pursued.

INDUSTRIAL COLLABORATION

Much of the work since 1990 has been in collaboration with private industry. Progress certainly would have been much slower than it has been without the help of a number of scientists in biotechnology companies. Unfortunately, because of the perceived lack of a commercial market for transmission-blocking vaccine and my ability to convincingly articulate the importance of such a vaccine, most of these scientists have worked without the complete support of their administrative officers. Because it may be counterproductive to identify individuals and companies involved, only the expression or delivery systems used will be identified.

Live Delivery Systems

A number of live vector delivery systems have been tested as vehicles for eliciting transmission-blocking antibodies. Because of the success with the WR strain of vaccinia virus, a significant effort was made to develop an attenuated strain of recombinant vaccinia virus (NYVAC-Pf7, see Reference 11) that expressed Pfs25. A corporate decision was made to pursue a highly attenuated vaccinia strain that expressed seven different malaria parasite antigens. Preliminary evidence suggests that, although experimental laboratory animals vaccinated with this recombinant vaccinia virus elicited antibodies that recognize Pfs25,[11] sera from these animals do not reproducibly block infectivity to mosquitoes by the membrane-feeding transmission-blocking assay (unpublished data).

A particularly promising approach using live vectors to elicit protective antibody responses has been to prime with the live virus and boost with a conventional subunit vaccine. Studies currently in progress in which experimental laboratory animals receive a series of vaccinations with recombinant attenuated vaccinia virus and then subsequently receive a booster injection with subunit vaccine suggest that this approach elicits high-titer transmission-blocking antibodies (unpublished data). Further studies will be necessary to determine whether this is a reproducible and feasible approach to elicit transmission-blocking activity in humans.

In addition to recombinant vaccinia virus, recombinant adenovirus, *Salmonella*, and naked DNA that encode Pfs25 have been prepared and studied in animals that have received multiple injections of the single delivery system or in a prime-boost format with subunit Pfs25 vaccine (unpublished data). To date, no experimental laboratory animal has developed transmission-blocking antibodies by either approach.

Subunit Approaches

A more traditional subunit vaccine approach has been developed in parallel to the live delivery systems described above. Besides the bacterial expression system that has yet to yield immunogenic transmission-blocking protein, I have pursued expression of recombinant Pfs25 in three eukaryotic systems: yeast, baculovirus-infected insect cells, and mammalian cells. Only in yeast have sufficient quantities of purified recombinant protein been made for human trials. The yeast system used to express Pfs25, first developed by a leading biotechnology company[12] and subsequently refined in my laboratory with help from another leading biotechnology corporation,[13] are much easier to use and scale-up than the cell culture systems required for baculovirus and mammalian cells. Therefore, the inability to manufacture suitable quantities of recombinant protein in baculovirus-infected insect cells or mammalian cells probably reflects more my lack of expertise and lack of a commercial biotechnology company partner rather than deficiencies in the expression system. Nevertheless, in my experience, recombinant yeast expression has provided a ready means of producing recombinant immunogens for clinical testing. It is still the system I would use first to explore recombinant protein expression if bacterial expression fails to produce biologically active protein.

Good Manufacturing Practices

The transition from producing research-grade material for preliminary animal studies to manufacturing clinical-grade material for preclinical and clinical studies can be expensive and problematic. Scale-up can take a rather nonlinear course and post-production process development is often very time-consuming. The resulting clinical-grade material may have markedly different biological activity from that produced on the benchtop. All of these difficulties have played out in the manufacture of subunit clinical-grade Pfs25.

The original yeast construct, Pfs25-B, looked extremely promising in animal studies.[12,14] Rapid progress was made in studies in primates with a delivery system (alum) suitable for use in humans. However, after a mutual decision by the biotechnology company and the funding agencies to move this product into GMP production, progress slowed and eventually ceased. In part, the problems that arose were programmatic; however, a marked decrease in yields on scale-up, aggregation, and degradation of the recombinant protein during post-production process development and the fact that, unlike the research-grade material, the clinical-grade material produced failed to reproducibly elicit transmission-blocking activity in mice finally led to the suspension of further development of Pfs25-B.

A substantial investment of time and resources were then required to redesign the manufacturing process, which led to the development of essentially a new product referred to as TBV25H[13] that was ultimately manufactured by a different biotechnology company. Because it was basically a philanthropic venture by the commercial partner, the project seemed imperiled by a number of legal questions that arose, particularly with regard to techniques used in the purification process. Once these were resolved, other questions arose as to the scalability of the final gel filtration step employed. Because size exclusion chromatography requires relatively large volumes of resin and has limitations in loading volumes to provide adequate resolution, and in addition is a diluting rather than a concentrating step, scale-up can be problematic. Since no adequate alternative was found before the assigned production date, the final post production process of the bulk antigen included a gel filtration step. Even if Phase I and II testing of a formulation of this bulk antigen looked promising, undoubtedly further work in post-production process development would be required before Phase III trials of the vaccine could be considered.

PRECLINICAL AND HUMAN TRIALS

To the uninitiated, the record-keeping/documentation required to comply with Good Laboratory Practices as described in the Code of Federal Regulations seemed at times to be all-consuming. Because of limited funds, my laboratory performed some of the preclinical studies to support the Investigational New Drug application. These studies required the *de novo* preparation of a number of standard operating procedure documents for a number of routine assays (e.g., SDS-PAGE, ELISA, and protein concentration) as well as assays unique to the development of transmission-blocking vaccine (e.g., production of infectious gametocytes, transmission-blocking assays, potency assays for transmission-blocking vaccines). For a basic research laboratory not accustomed to GLP records, the rather onerous task of generating these documents often met with resistance by the laboratory staff. The task took in excess of two years to complete, in part because of technical difficulties in manufacturing the bulk antigen as described above.

Two clinical trials of transmission-blocking vaccines have been undertaken (unpublished data). Because both are still in progress or have yet to be published, the description of the results from those trials that follows is incomplete. The first was a Phase I trial of a highly attenuated vaccinia virus that encodes seven malaria parasite antigens, one of which is Pfs25. Data from preclinical studies in experimental laboratory animals indicated that Pfs25 was immunogenic; however, complete transmission-blocking activity in animals or humans was not observed (unpublished data).

The second human trial was of the yeast-produced alum-adsorbed TBV25H construct described above. Preliminary analysis of the sera collected from the human volunteers after the third dose of TBV25H/alum vaccine suggests that the formulation is immunogenic. Because one of seven volunteers who received the test vaccine experienced a significant local adverse event after the third dose of vaccine, whether the alum-adsorbed delivery system is without significant safety problems still remains to be determined.

THE FUTURE: PRIME-BOOST AND FUSION PROTEINS

Perhaps the most promising transmission-blocking results in experimental laboratory animals vaccinated with clinical-grade vaccine are in prime-boost studies in which animals received NYVAC-Pf7 three times[11] and then were further boosted by a single injection of TBV25H/alum (unpublished data). Clearly further studies will be necessary; however, to prove the principle that transmission-blocking antibodies can be elicited at all in humans, the same prime-boost strategy is currently being tested in a Phase I human trial. Although deployment of such a vaccine regimen may present some logistical problems in developing countries, the strategy may still prove extremely useful in "proving the principle" in Phase II/III field studies of transmission-blocking vaccine efficacy.

A second-generation subunit yeast-produced vaccine, TBV25-28, consisting of a fusion protein of two sexual-stage antigens, Pfs25 and Pfs28, looks particularly promising (unpublished data). The vaccine appears to be substantially more potent, in that fewer doses (two rather than three) were required, less antigen (approximately one-fifth as much) was used, and a longer immune response (greater than 4 months after the third vaccination) has been observed for an alum-adsorbed preparation in mice. It has been more difficult to elicit transmission-blocking in rabbits than in mice. Therefore, a pivotal study in lagomorphs is in progress to determine whether this second-generation vaccine can elicit transmission-blocking activity. If so, then human Phase I testing of this new vaccine seems warranted.

SUMMARY

It is important to emphasize that the development of malaria vaccine, particularly transmission-blocking vaccine, is still in its early days. Despite impressive strides in our understanding of malaria parasites, years of work and unforeseen hurdles still face us. Still, vaccines hold the greatest hope of a long-term solution to this ancient scourge; thus, despite "diminishing public funds, fragmented public sector efforts, and limited interest within the vaccine industry,"[15] scientific optimism in malaria vaccines continues to burgeon as it must if we are to succeed.

ACKNOWLEDGMENTS

The opinions of the author do not necessarily reflect those of NIAID, NIH, DHHS, or the numerous collaborators who contributed to the work described herein. The author gratefully acknowledges WHO/TDR/IMMAL (particularly Drs. T. Godal, P. Reeve, and H. Engers) and LPD-LMR/DIR/NIAID (particularly Drs. F.A. Neva, L.H. Miller, J. Gallin, and T. Kindt) for supporting this research for more than a decade.

REFERENCES

1. Laveran, A. 1880. Note sur un noveau parasite trouve dans le sang du plusiers malades atteints de fievre palustre. *Bull. Acad. Natl. Med. Paris* 9:1235.
2. Harrison, G.A. 1978. In: *Mosquitoes, Malaria and Man: A History of Hostilities Since 1980*, p. 11, E.P. Dutton, NY.
3. Ross, R. 1898. Report on the cultivation of proteosoma, Labbe, in grey mosquitoes. *Indian Med. Gaz.* 33:401–408.
4. MacCallum, W.G. 1898. On the haematozoan infections in birds. *J. Exp. Med.* 3:117–136.

5. Gwadz, R.W. 1976. Malaria: Successful immunization against sexual stage of *Plasmodium gallinaceum. Science* 193:1150–1151.
6. Carter, R. and D.H. Chen. 1976. Malaria transmission blocked by immunisation with gametes of the malaria parasite. *Nature* (London) 263:57–60.
7. Vermeulen, A.N., T. Ponnudurai, P.J.A. Beckers, J. Verhave, M.A. Smits, and J.H.E. Meuwissen. 1985. Sequential expression of antigens on sexual stage of plasmodium falciparum accessible to transmission-blocking antibodies in the mosquito. *J. Exp. Med.* 162:1460–1476.
8. Kaslow D.C., I.A. Quakyï, C.Syin, M.G. Raum, D.B. Keister, J.E. Coligan, T.F. McCutchan, and L.H. Miller. 1988. A vaccine candidate from the sexual stage of human malaria that contains EGF-like domains. *Nature* (London) 333:74–76.
9. Kaslow, D.C., I.C. Bathhurst, T. Lensen, T. Ponnudurai, P.J. Barr, and D.B. Keister. 1994. Saccharomyces cerevisiae recombinant Pfs25 adsorbed to alum elicits antibodies that block transmission of *Plasmodium falciparum. Infect. Immun.* 62:5576–5580.
10. Kaslow, D.C., S.N. Isaacs, I.A. Quakyi, R.W. Gwadz, B. Moss, and D.B. Keister. 1991. Induction of *Plasmodium falciparum* transmission-blocking antibodies by recombinant vaccinia virus. *Science* 252:1310–1313.
11. Tine, J.A., D.E. Lanar, D.M. Smith, B.T. Wellde, P. Schultheiss, L.A. Ware, E.B. Kauffman, R.A. Wirtz, C. De Taisne, G.S.N. Hui, S.P. Chang, P. Church, M.R. Hollingdale, D.C. Kaslow, S. Hoffman, K.P. Guito, W.R. Ballou, J.C. Sadoff, and E. Paoletti. 1996. NYVAC-Pf7: A Poxvirus-vectored, multiantigen, multistage vaccine candidate for *Plasmodium faliparum* malaria. *Infect. Immun.* 64:9:3833–3844.
12. Barr, P.J., K.M. Green, H.L. Gibson, I.C. Bathurst, I.A. Quakyi, and D.C. Kaslow. 1991. Recombinant Pfs25 protein of *Plasmodium falciparum* elicits malaria transmission-blocking immunity in experimental animals. *J. Exp. Med.* 174:1203–1208.
13. Kaslow, D.C. and J. Shiloach. 1994. Production, purification and immunogenicity of a malaria transmission-blocking vaccine candidate: TBV25-H expressed in yeast and purified using nickel-NTA agarose. *Bio/Technology* 12:494–499.
14. Kaslow, D.C., I.C. Bathurst, D.B. Keister, G.H. Campbell, S. Adams, C.L. Morris, J.S. Sullivan, P.J. Barr, and W.E. Collins. 1993. Safety, immunogenicity and *in vitro* efficacy of a muramyl tripeptide-based malaria transmission-blocking vaccine in an *Aotus nancymai* monkey model. *Vaccine Res.* 2:95–103.
15. *An Institute of Medicine Report.* 1996. *Vaccines Against Malaria: Hope in a Gathering Storm.* Russel, P.K. and C.P. Howson (Eds.) National Academy Press, Washington, DC.

The Victories and Vexations of Vaccine Production— The Varicella Vaccine

Michiaki Takahashi

BACKGROUND

Attenuation of Measles Virus, Rubella Virus, and Polioviruses

In 1959–1962, I worked on development of attenuated live measles vaccines in the laboratory of Professor Yoshiomi Okuno, in Osaka. This virus was attenuated by passage in the amniotic cavity and chorioallantoic membrane of developing chick embryos.[1,2] In addition to my work with measles virus, I was also asked to do a study on adaptation of poliovirus types 1 and 3 to chick embryo cells. As is well known, poliovirus type 2 grows well in developing chick embryo cells, but types 1 and 3 do not. I attempted to adapt these viruses—particularly type 3—to chick embryo cells by alternate passage in chick embryo cells and monkey kidney cells.[3,4] The attempt finally failed: no continuous growth of poliovirus type 3 took place in chick embryo cells. However, after several alternate passages the virus was found to be thermosensitive (i.e., the titer of the passaged virus was lower at 34°C than at 39°C whereas titers of the original strain were comparable at these two temperatures) and to be less neurovirulent when inoculated into the thalamus of monkeys.

From these studies, I learned that passage in foreign-species cells is a convenient and effective means by which to attenuate viruses for use in live virus vaccines.

Malignant Transformation of Cultured Cells with Human Adenovirus and Herpes Simplex Virus

I had long been interested in the possible causative relationship of human viruses to human cancer. In 1962, tumor formation by adenovirus type 12 was reported in newborn hamsters.[5] Stimulated by that finding, I started *in vitro* transformation experiments with adenovirus type 12; no viral growth or lytic viral infection was detectable in inoculated hamster embryo cells. In contrast, adenovirus type 5, which was classified as a nontumorigenic virus, cause lytic infection in hamster embryo cells. Both viruses are lytic to human embryo cells. Thus we tried to obtain conditional lethal mutants of adenovirus type 5 and to ascertain whether such mutants could—like adenovirus type 12—cause the transformation of hamster cells. We obtained both temperature-sensitive mutants, which grew at 38.5°C, and host-dependent mutants which caused lytic infection in human but not hamster embryo cells.[6] Using an established hamster embryo cell line (Nil cells) that, unlike primary cultured cells, is readily transformed, we observed malignant transformation with both mutants.[7,8] However, we detected no transformation of human embryo cells with these mutants. This finding was consistent with the lack of evidence of a causative relationship of human adenoviruses with human cancer.

In 1971, Duff and Rapp reported that hamster embryo fibroblasts were transformed with ultraviolet-irradiated herpes simplex virus type 2 (HSV-2).[9] We found their work interesting and attempted to transform hamster cells with temperature-sensitive mutants at a nonpermissive temperature. Approximately 2,700 clones of HSV-2 from mutagenized stock virus and were isolated at 32°C, and 42 clones found to be nonpermissive at 38.5°C were examined for the ability to transform hamster and human embryo cells at 38.5°C. Hamster embryo cells were transformed by three mutants.[10] Transient transformation of human embryo fibroblasts was documented with one mutant, but resulted in the failure of serial passage of the cells so that the finding was not reproducible. Later on, we attempted repeatedly to transform human embryo fibroblasts with ultraviolet-irradiated human HSV-2, but were unsuccessful.

Through these experiments, I became convinced that human adenovirus and HSV, although known to induce malignant transformation of hamster and rat embryo fibroblasts (i.e., foreign-species cells) are related little—if at all—to oncogenesis in human cells (i.e., indigenous cells).

MOTIVATIONS FOR AND PROBLEMS IN THE DEVELOPMENT OF A LIVE VARICELLA VACCINE

Chickenpox is usually a mild illness but occasionally manifests as a severe disease in children. After a member of my family had severe chickenpox in 1964, with high fever and widespread rashes lasting for 3 days, I began to consider how this disease might be prevented by vaccination. Since I knew that live vaccines induced solid immunity against diseases such as measles and polio, my thought from the beginning of the study was to develop a live attenuated varicella vaccine.

Two major problems had to be considered. The first was the possible oncogenicity of varicella-zoster virus (VZV), which is a herpesvirus. Through the experiences described above, I had been convinced that HSV is either minimally or totally unrelated to malignancy in human cells. Although it was difficult to rule out VZV as a cause of malignancies, VZV had never been linked to any form of cancer. After my studies, Gelb et al.[11] reported that their fresh VZV isolates transformed hamster embryo cells morphologically, but they later reported[12] that this observation was not reproducible. Thus, even *in vitro*, it seemed unlikely that VZV could induce malignant change. The second problem was the possibility that live varicella vaccine virus would become latent, perhaps resulting in the later development of zoster. I presumed that attenuated virus would have less capacity than wild-type virus to replicate in humans and thus to become latent. In addition, I expected that symptoms of zoster caused by attenuated virus might be less severe than those of disease caused by wild-type viruses. Thus I thought that these two issues were not obstacles to the development of a live varicella vaccine.

DIFFICULTIES IN PREPARING "CELL-FREE" VZV

Since the earliest studies on *in vitro* propagation of VZV, it has been recognized that virus produced in cell cultures remains strongly cell associated; the inability to obtain cell-free infectious virus has hampered biological and immunological studies on this virus. Caunt[13] and Caunt and Taylor-Robinson[14] showed that infectious VZV could be isolated in a cell-free state following ultrasonic disruption of infected primary human thyroid cells. Shortly thereafter, Brunell[15] reported the isolation of cell-free virus from infected human embryo lung fibroblasts. Referring to those papers, we undertook studies to identify a suitable method for the isolation of cell-free virus from infected cultures and the composition of a suspending medium that would keep the infectivity of the virus as stable as possible. We reasoned that the following procedures would be likely to yield high-titered cell-free virus from infected cells: (1) use of cultured cells in the growth phase for inoculation of virus; (2) high-input multiplicity, with infected cells (rather than cell-free virus) used for inoculation because of the difficulty of obtaining a sufficient dose of cell-free virus; and (3) harvesting of the infected cell monolayer (by treatment with EDTA) before the appearance of advanced cytopathic changes with subsequent preparation of the infected cell suspension.

Because VZV is highly heat-labile, particular caution was required in the selection of a suspending medium that would preserve its infectivity. After comparison of various media, simple phosphate buffered saline (Ca^{++}, Mg^{++} free) was selected as the most suitable with sucrose (final concentration, 5%), sodium glutamate (0.1%), and fetal calf serum (10%, or 2.5% gelatin hydrolysate in case of vaccine preparation).[16] With this medium, the decrease in infectivity during storage at –70°C was minimal; in fact, no decrease was detectable after storage for 1 year.

PRIMARY ISOLATION OF VACCINE VIRUS

Fluid was taken from the vesicles of a 3-year-old boy who had typical chickenpox but was otherwise healthy. The fluid was stored at –70°C until it was inoculated onto primary cultures of human embryonic lung (HEL) cells. At a temperature of 34°C, characteristic foci appeared after 7–10 days. The virus was designated as the Oka strain since this was the surname of the boy from whose vesicular fluid it was derived.[17]

RATIONALE FOR AND DESIGN OF A LIVE VARICELLA VACCINE

VZV spreads from cell to cell, forming distinct foci that are visible by microscopy even in unstained cell cultures and are clearly seen after methylene blue or fluorescent antibody staining. Cell-mediated immunity seems essential—or at least as important as humoral immunity—in preventing the spread of VZV *in vivo*. Since inactivated or subunit viral antigens are usually weak inducers of cell-mediated immunity, we reasoned that a live vaccine might be most useful for the prevention of varicella.

It had been very difficult to demonstrate the pathogenicity of VZV in experimental animals. Therefore, we anticipated that the attenuation of VZV would be proven only by extensive clinical trials, and that testing of only a limited number of candidate strains would be feasible. The classical (empirical) method of attenuation, as described previously, was used. Of the various kinds of nonprimate cultured cells tested for susceptibility to infection with the Oka strain of VZV, only guinea pig embryo fibroblasts (GPEF) were found susceptible. Of the several VZV strains tested, the Oka strain grew best in GPEF, when initially used as the substrate cells for vaccine studies. Until the early 1970s, nonprimate primary cell cultures were preferred to continuous human cell lines for vaccine production. In 1969, at a meeting on rubella vaccine at the U.S. National Institutes of Health, excellent data was presented by Plotkin et al.[18] concerning RA27/3 vaccine prepared in human diploid cells (WI-38). However, opposition to the diploid cell substrate was expressed by Sabin[19,20] who stressed the theoretical risk of human retrovirus in human diploid cells. On the basis of these objections, the first live rubella vaccines approved by U.S. federal authorities in 1969–1970 were prepared in duck embryo cell culture (HPV 77 strain),[21] dog kidney cells (HPV 77 strain),[22] and rabbit kidney cell culture (Cendehill strain).[23] Soon thereafter the RA/27 human diploid fibroblast vaccine was licensed in Europe.[24]

On the basis of these results, we conducted further experiments aimed at the development of a live varicella vaccine. After the 11th passage of Oka-strain virus in HEL cells at 34°C, infected cells were trypsinized and inoculated onto GPEF. Characteristic CPE appeared in a few days, and the transfer of infected cells was repeated. Cell-free virus (1,000 to 2,000 PFU/mL) was extracted from infected cells by sonication. Passaged virus was identified as VZV by hemotoxylin-eosin staining, fluorescent antibody staining, and the neutralization test using HEL cells. Oka-strain VZV thus passaged 11 times in HEL cells and 6 times in GPEF was slightly more thermosensitive at 39°C than wild-type viruses and exhibited a greater capacity for growth in GPEF than the original or other wild-type strains. The biological and biophysical properties of this vaccine virus were described in detail in later reports.[25,26] Safety testing of the vaccine revealed a lack of pathogenicity (including intracerebral effects) in small nonprimate mammals and monkeys. The absence of C-type particles and of latent viruses was also confirmed morphologically and biochemically.

EARLY CLINICAL TRIALS: VACCINATION OF HEALTHY AND HOSPITALIZED CHILDREN

With the informed consent of their parents, healthy children who were living at home and had no history of varicella received various doses of Oka-strain virus passaged six times in GPEF.[17] A dose of 500 PFU elicited seroconversion in 19 of 20 children. Even at a dose of 200 PFU, an antibody response was detected in 11 of 12 children. No symptoms due to vaccination were detected in these children. In short, sixth-passage Oka-strain virus in GPEF was well tolerated and immunogenic.

The first clinical trial of the vaccine in hospitalized children was undertaken in an effort to terminate the spread of varicella among children with no history of the disease.[17] In the hospital where the trial was conducted, chickenpox had frequently spread in the children's ward with severe cases on some occasions. In this protocol, children with no history of varicella were vaccinated immediately after the occurrence of a case of varicella. These children suffered from conditions including nephrotic syndrome, nephritis, purulent meningitis, and hepatitis. Twelve children had been receiving corticosteroid therapy. An antibody response was documented in all of the vaccinated children; within 10–14 days after vaccination six children developed a mild fever, and two of the six developed a mild rash. It was uncertain whether these reactions were due to vaccination or to naturally acquired infection modified by vaccination. No other clinical reactions or abnormalities of the blood or the urine were detected. Thus on this ward, the spread of varicella infection was prevented except in one case: a child who was not vaccinated because his mother mistakenly believed that he already had varicella became severely ill. This study offered the first proof that the Oka vaccine was well tolerated by patients receiving immunosuppressive therapy and stirred hopes that this vaccine would prove practical for the prevention of varicella.

CLINICAL TRIALS WITH VACCINES PREPARED IN HUMAN DIPLOID CELLS

VZV yield from GPEF cells was considerably lower than that from human embryo fibroblasts. In addition, the level of viral infectivity was found to decrease to approximately one-third of the original level during lyophilization. Thus, cells that would yield more virus were sought. Because the human diploid cell line WI-38 had been widely used for vaccine production, we decided to cultivate the Oka strain of VZV in WI-38 cells. After 12 passages in GPEF, the virus was passaged several times in WI-38 cells. The virus thus obtained was subjected to the same safety testing described previously and was evaluated in clinical trials. When a shortage in the supply of WI-38 cells became a concern,[20] MRC-5 cells[27] were assessed. A master seed lot was prepared at the second passage level in MRC-5 cells after three passages in WI-38 cells, and vaccines were subsequently produced exclusively in MRC-5 cells.

In an examination of its protective efficacy, the resulting vaccine was given to susceptible household contacts immediately after exposure to varicella.[28] Twenty-six contacts (all children) from 21 families were vaccinated, mostly within three days after exposure to the index cases. None of the vaccinated children developed symptoms of varicella. In contrast, all 19 unvaccinated contacts (from 15 families), exhibited typical varicella symptoms 10–20 days after the onset of the index cases. In three families where one sibling contact received vaccine and the other did not, none of the vaccinated children developed symptoms, whereas all unvaccinated controls exhibited typical symptoms. In general, the antibody titers after clinical varicella were 8–10 times higher than those after immunization. This study clearly demonstrated that vaccination soon after exposure was protective against clinical varicella.

In another clinical study,[29] immunized children on a hospital ward were protected despite subsequent exposures to natural varicella and herpes zoster during the nine months after vaccination. After two years of follow-up of 179 vaccinated children including 54 children who had been receiving steroid therapy, 50 (98%) remained seropositive in the neutralization test, and only one of 13 household contacts of cases manifested mild varicella (10 vesicles but no fever).[30] In an institution for children less than two years old, prompt vaccination had a similar protective effect.[31] Varicella developed in an 11-month-old infant on a ward for 86 children. A total of 33 children over 11 months of age were vaccinated; 43 children less than 11 months of age were not vaccinated, partly because they were expected to still possess maternal antibody. A small viral dose (80 PFU) was used for immunization. Of the vaccinated group, 8 developed a mild rash and 1 of these 8 had a mild fever (less than 38°C) in 2–4 weeks after vaccination. In contrast, typical varicella developed in all 43 unvaccinated children during the 10 weeks after onset of the index case. Symptoms were severe in 16 cases, with confluent vesicles and high fever; after recovery, scars remained in 13 of these 16 cases. These results suggested that vaccination with as little as 80 PFU frequently stopped the spread of varicella among children in close contact with one another.

VACCINATION OF CHILDREN WITH MALIGNANT DISEASES

In the first vaccination trials in children with malignant diseases with virus doses of 200, 500, or 1,500 PFU, chemotherapy was suspended for 1 week before and 1 week after vaccination.[25,32,33] Of 12 immunized children with acute lymphocytic leukemia (ALL), 10 who had been in remission for 6 months or less, 1 for 9 months, and 1 for 48 months, 4 had fewer than 3,000 white blood cells/mm,[3] but most had positive skin-test reactions with dinitrochlorobenzene, purified protein derivative, or phytohemagglutinin. Three of the 12 children developed a mild rash, with 13 (with 1,500 PFU), 30, and 25 (with 200 PFU) papular or incomplete vesicles, respectively; one child had a fever (39°C) for 1 day about 3 weeks after vaccination. These results offered hope that a live varicella vaccine could be administered, with some precautions, to high-risk children.

VIEWPOINTS REGARDING LIVE VARICELLA VACCINE AFTER INITIAL CLINICAL TRIALS

The various viewpoints expressed regarding live varicella vaccine in 1977, after these early clinical trials, included a cogent commentary by Brunell[34] whose main points were as follows:

1. The vaccine itself may cause zoster; however, it will take decades to find out whether or not this is the case.
2. Unfortunately, markers predictive of the behavior of a given strain of VZV with respect to causing zoster have not been identified.
3. Immunity after vaccination may not be as long-lasting as that after natural infection; thus, vaccination may enhance the risk of the relatively severe disease that frequently follows in adulthood.
4. Since naturally occurring varicella can be severe or even fatal in immunocompromised children who are receiving steroids for various chronic conditions and in patients with leukemia, it is not clear whether a live varicella would protect these children or cause serious disease, and it will be hard to find out.

Albert Sabin[35] presented the following views on the matters discussed by Dr. Brunell:

1. There is a high probability that live varicella vaccine virus will cause zoster infrequently: the absence of lesions and clinical manifestations in vaccinated children indicates that there is only limited viral multiplications and dissemination in the body and thus the potential for only limited (or no) invasion of sensory ganglia.
2. The lack of markers for zoster is not a contraindication for the testing or use of live varicella vaccine; live measles and rubella vaccines are being used in the absence of disease-specific markers.
3. The duration of immunity following the injection of a varicella vaccine is, of course, important, but it can be determined.

Stanley Plotkin[36] expressed a viewpoint in opposition to Dr. Brunell's, emphasizing that authority, however well-meaning, should not stand in the way of gathering data as long as the consequences are weighed at each step. While stressing the need for caution, Brunell[37] replied that he wholeheartedly supported research that would increase the understanding of virus latency. Drs. C. Henry

Kempe and Anne Gershon[38] stated that although varicella vaccine might result in either an increase or a decrease in latency, there was a real possibility of the latter, and only long-term studies of vaccinees would provide an answer. They reminded readers that in any experimental endeavor involving human beings, the risk/benefit ratio is of immense importance. On the basis of the available data, they concluded that the potential benefits of varicella vaccine might well outweigh the potential dangers, particularly in high-risk persons.

Thus, in 1977, conflicting opinions were exchanged among several distinguished scientists interested in viral vaccines. Most of them favored continued work on a live varicella vaccine, including further studies on the latency of vaccine virus and the likelihood of subsequent zoster.

CLINICAL VACCINE TRIALS IN THE
UNITED STATES AND EUROPE

In February 1979, a workshop on VZV was held at the U.S. National Institutes of Health. The main topic of discussion was whether or not varicella vaccine should be evaluated in clinical trials involving high-risk children. Dr. Saul Krugman referred to good short-term results with the vaccine, which he thought deserved to be tested. An NIH Collaborative Study Group was organized (Chief, Dr. Anne Gershon of New York University and later of Columbia University), and clinical trials were started with Oka-strain live varicella vaccine produced by Merck Research Laboratories, West Point, New York. Many excellent investigations[39–42] were conducted by that group, including clinical reactogenicity, the frequency of household transmission from vaccinated acute leukemic children with rash, and the persistence of immunity. Other study groups also conducted clinical trials, most of which yielded favorable results.[43–47] In Europe, clinical trials were conducted with Oka-strain varicella vaccine prepared by SmithKline RIT. In 1983, an Expert-Committee meeting (Chief, Dr. F.T. Perkins) was held at the World Health Organization in Geneva to prepare a manuscript entitled "Requirements for the Live Varicella Vaccine." The resulting document was circulated and reviewed by authorities around the world and finally was published in 1985.[48] Meanwhile, in 1984, the Oka-strain live varicella vaccine produced by SmithKline RIT was licensed for administration to high-risk children in Austria, Belgium, Federal Republic of Germany, Ireland, Luxembourg, Portugal, Switzerland, and the U.K. In November 1984, a symposium on active immunization against varicella was held in Munich. The papers presented at this meeting were published in the following year.[49–61] I was personally encouraged to read in this publication Dr. Plotkin's statement[60] that, despite a few questions, varicella vaccine appears to have a bright future, and that the work of Professor M. Takahashi, conducted over more than 10 years, deserved praise, as he has persevered in the face of criticism, bringing medical science to the point where we can contemplate the conquest of another widespread human disease.

DETECTION OF VIREMIA BEFORE AND AT THE ONSET OF CHICKENPOX: HERPES ZOSTER AND VARICELLA VACCINE

It had been extremely difficult to isolate VZV from blood and secretions except in severely immunocompromised individuals. Moreover, no evidence of hematogenous spread of VZV in normal children had been documented. Nevertheless, our colleagues Ozaki et al.[62] succeeded in isolating VZV at a high rate from the mononucleocytes of otherwise healthy children with chickenpox. Within 5 hours of blood collection, these authors inoculated the mononucleocyte fraction onto human embryo cell cultures at a ratio of 1:1. When no cytopathic effect was observed, the cells were trypsinized and transferred into a new monolayer. Virus was isolated from all of 4 children from whom blood samples were taken on day 1 of illness. It was also possible to isolate VZV during the incubation period (i.e., 1–5 days before the appearance of rash).[63,64] In contrast, no VZV could be recovered from any of 28 children 4–14 days after vaccination (before the appearance of neutralizing antibody) even at a high dose of 5,000 PFU.[65] These findings had significant implications for the pathogenesis of chickenpox and zoster and particularly for the issue of latency of vaccine virus in immunized individuals.

The current views of the immunopathogenesis of chickenpox and VZV to viremia are as follows: The initial site of infection with wild-type VZV may be the conjunctiva, the upper respiratory tract, or both. Conjunctival and respiratory infections have been demonstrated in a guinea pig model.[66] The virus then replicates at a local site, probably in regional cervical lymph tissues. Lymph node infection by VZV has been confirmed by the detection of VZV antigen in lymph nodes at autopsy in a fatal zoster case.[67] The incubation period of natural varicella infection is usually 14 or 15 days, and the duration of local viral replication is estimated to be 4–6 days.[68] This estimate is justified by the effect of the immediate inoculation of household contact with live varicella vaccine.[28] The vaccine virus bypasses the initial replication site of wild-type VZV, thereby inducing immunity so early as to prevent subsequent replication of wild-type virus.

Secondary viremia may follow after replication in some or all of the above-mentioned sites. Of greater magnitude than the primary viremia, secondary viremia delivers virus to the skin, thus causing the appearance of rash. The occurrence of secondary viremia in natural varicella infection has been demonstrated by the isolation of VZV from mononuclear cells of immunocompetent patients, as described above.[62–65] In contrast, no VZV was recovered from vaccinees.[65]

These results suggested that the replication of vaccine virus in susceptible viscera is of lesser magnitude than that of wild-type VZV but is sufficient to induce an immune response mainly in regional lymph nodes. Thus it seemed that viremia was proportional to the virulence of the virus involved and that it was correlated with the appearance of a rash in otherwise-healthy children.

It has generally been believed that VZV in the skin vesicles travels up the sensory nerves to posterior ganglia, where it persists; this seems to be the major route of virus migration. Hope-Simpson[69] noted that the pattern of incidence of zoster on individual sensory ganglia is similar to the distribution of the rash in chickenpox and may bear a direct relationship to it. This observation may explain why sensory ganglia, and not motor ganglia, are selected for viral lodgement. As mentioned at the beginning of this chapter, a major question about live varicella vaccine had been whether the vaccine virus becomes latent and causes the later development of zoster. Since zoster is relatively uncommon in healthy children, long-term follow-up of vaccinated healthy children was required to answer this question definitively. However, since children with acute leukemia tend to develop zoster soon after natural infection, it was assumed that careful observation of the incidence of zoster in vaccinated children with ALL would yield valuable insight. Thus, vaccinated leukemic children were followed closely for the development of zoster and compared with leukemic children who had had natural varicella.

In one study group in Japan, the incidence of zoster among vaccinated and naturally infected children were 15.4% (n = 52) and 17.5% (n = 43), respectively;[70] in another, the rates were 9.1% (n = 44), and 21.6% (n = 37), respectively.[71] Clinical symptoms in vaccinated children were usually mild and untroublesome, while in the naturally infected children in the latter study group, one had moderate and the others had severe symptoms. Some VZV isolates from cases of zoster that developed in vaccinated patients with ALL were shown to be derived from vaccine virus.[26] However, all of the individuals studied had underlying acute leukemia, and person-to-person variation in their physical condition might have complicated precise comparison of the incidence of zoster in the two groups. As I contemplated how we could obtain more definitive evidence on the incidence of zoster after vaccination of children with acute leukemia, it occurred to me that the incidence of zoster should be followed in two groups of children with acute leukemia—one group who developed rash after vaccination and one who did not. As noted previously, the major route by which VZV reaches ganglia seems to be along peripheral nerves from vesicles. If the incidence of zoster were found to be higher among children who developed a rash after vaccination than among those who did not, then we should be able to predict whether latent infection of vaccine virus will occur in immunized children depending on their reaction to the vaccine.

Thus, we made this comparison, and the results shed more light on the latency of vaccine virus in vaccine recipients. In a retrospective follow-up study of children with acute leukemia, zoster occurred far more frequently in those who developed a rash after vaccination (17.1% or 3.13 cases per 100 person-years; n=70) than in those without rash (2.4%, or 0.46 cases per 100 persons-years; n=250).[73-76] These figures suggested that an absence of rash after vaccination is directly correlated with a low incidence of zoster, which in turn indicates that the incidence of zoster is lower among vaccine recipients than among children who have natural varicella.

In 1986, live varicella vaccine produced by the Research Foundation for Microbial Diseases of Osaka University was licensed in Japan for use in high risk children and for optional use in children at standard risk. In 1988, a live varicella vaccine of the Oka strain similar to that licensed in Japan, was licensed in Korea.

FURTHER CLINICAL STUDIES IN THE UNITED STATES PARTICULARLY ON THE INCIDENCE OF ZOSTER AFTER VACCINATION

Studies from the United States have indicated more clearly that the incidence of zoster after vaccination of leukemic children is lower than that after natural infection. Brunell et al.[72] reported that 19 of 26 children with acute leukemia who had natural varicella developed zoster, while none of 48 vaccinees did. With adjustment for the duration of observation and exclusion of vaccinees who failed to have a sustained antibody response or to develop chickenpox, the risk of zoster was still lower among vaccinees ($P = 0.017$). The investigators concluded that there is no reason to suspect that recipients of varicella vaccine are more likely to develop zoster than children who have varicella.

One comparative study included 84 matched pairs of U.S. children with underlying acute leukemia. Zoster developed in three (3.6%) of the 84 vaccinated subjects during 2,936 months of observation (an incidence of 1.23 cases per 100 person-years) and in 11 (13.1%) of the 84 naturally infected subjects during 4,245 months of observation (an incidence of 3.11 cases per 100 person-years).[40]

Further studies by the NIAID Collaborative Study Group showed clearly that absence of rash is correlated with low incidence of zoster.[77] In their investigation of vaccinated children with acute leukemia who developed zoster, 11 had a rash due to VZV (a vaccine-associated rash in eight cases and breakthrough varicella in three). The two children in whom zoster developed without a VZV skin rash had zoster lesions at the site of vaccination. Of 268 vaccinated children with VZV rashes, 11 (4.1%) had zoster. In contrast, there were only two cases of zoster (0.7%) among the 280 vaccinated children with no VZV rash ($P = 0.02$ by Chi-squared test with continuity correction). The relative risk of zoster in the children who had had a VZV rash was 5.75 (95% confidence interval, 1.3–25.7).

Besides the main migration route (i.e., via the sensory nerve), there may be a minor hematogenous route of migration by virus to the ganglia.[78] As noted above, however, no viremia could be detected in healthy vaccine recipients. Therefore, whatever the route, it seems far less likely for the vaccine virus than for wild-type virus to become latent in the ganglia and cause subsequent zoster.

Given these results and current knowledge on the pathogenesis of herpes zoster, we can be convinced that immunization with live varicella vaccine would lead to a significant decrease in incidence of herpes zoster.

In 1995, Oka-strain live varicella vaccines (produced by Merck Research Laboratories and SmithKline Beecham, respectively) were licensed for healthy children in the United States and Europe.

ACKNOWLEDGMENTS

I am thankful for many collaborators in Japan, particularly Drs. Yoshizo Asano (Fujitagakuen Health University, Aichi), Hitoshi Kamiya (National Mie Hospital, Mie), Koichi Baba (Osaka University, Osaka) Kiyoshi Horiuchi (Jikei Medical School, Tokyo), Takao Ozaki (Showa Hospital, Aichi), Mikio Kimura (Tokai University, Kanagawa), Koichi Yamanishi (Osaka University, Osaka), Kimiyasu Shiraki (Toyama Medical School, Toyama), and Yasuhiko Hayakawa (Sendai City Health Institute, Miyagi).

I am also grateful to Drs. Philip A. Brunell, Anne A. Gershon, Stanley A. Plotkin, Stuart E. Starr, and Ann M. Arvin, and the late Drs. C. Henry Kempe, Saul Krugman, and Albert B. Sabin for their valuable criticism and encouragement during the course of our studies. I sincerely appreciate Dr. Lawrence C. Paoletti's careful and thoughtful review of the manuscript and the Brigham and Women's Hospital Editorial Service's detailed grammatical editing.

REFERENCES

1. Okuno, Y., T. Sugai, T. Fujita, T. Yamamura, K. Toyoshima, M. Takahashi, K. Nakamura, and N. Kunita. 1960. Studies on the prophylaxis of measles with attenuated living virus. II. Cultivation of measles virus isolated in tissue culture in developing chick egg. *Biken J.* 3:107–113.

2. Okuno, Y., M. Takahashi, M. Toyosima, T. Yamamura, T. Sugai, K. Nakamura, and N. Kunita. 1960. Studies on the prophylaxis of measles with attenuated living virus. III. Inoculation tests in man and monkeys with chick embryo passage measles virus. *Biken J.* 3:115–122.

3. Takahashi, M., S. Okabe, and M. Onaka. 1962. Studies on attenuation of polio virus (type 3, Saukett strain) by modified passage in developing chick embryo and chick cell culture. *Biken J.* 5:67–76.

4. Takahashi, M., S. Hamada, and S. Okabe. 1963. Characteristics of attenuated type 3 poliovirus obtained by alternate passage in chick cell cultures and monkey kidney cell culture. *Biken J.* 6:219–222.

5. Trentin, J.J., Y. Yabe, and G. Taylor. 1962. The quest for human cancer viruses. *Science* 137:835–849.

6. Takahashi, M. 1972. Isolation of conditional lethal mutants (temperature sensitive and host-dependent mutants) of adenovirus type 5. *Virology* 49:815–817.

7. Takahashi, M., Y. Minekawa, and K. Yamanishi. 1974. Transformation of a hamster embryo cell line (Nil) with a host-dependent mutant of adenovirus type 5. *Virology* 57:300–303.

8. Minekawa, Y., M. Ishibashi, H. Yasue, and M. Takahashi. 1976. Characterization of host-range and temperature sensitive mutants of adenovirus type 5

with particular regard to transformation of a hamster embryo cell line (Nil). *Virology* 71:97–110.

9. Duff, R. and F. Rapp. 1971. Oncogenic transformation of hamster cells after exposure to herpes simplex virus type 2. *Nature* 233:45–50.

10. Takahashi, M. and K. Yamanishi. 1974. Transformation of hamster and human embryo cells by temperature sensitive mutants of herpes simplex virus type 2. *Virology* 61:306–311.

11. Gelb, L., J.J. Huang, and W.J. Wellinghoff. 1980. Varicella-zoster virus transformation of hamster embryo cells. *J. Gen. Virol.* 51:171–177.

12. Gelb, L. and D. Dohner. 1984. Varicella-zoster virus-induced transformation of mammalian cells *in vitro*. *J. Invest Dermatol.* 83:77s–81s.

13. Caunt, A.E. 1963. Growth of varicella-zoster virus in human thyroid tissue cultures. *Lancet* 2:982–983.

14. Caunt, A.E. and Taylor-Robinson. 1964. Cell-free varicella-zoster virus in tissue culture. *J. Hyg.* (London) 62:413–424.

15. Brunell, P.A. 1967. Separation of infectious varicella-zoster virus from human embryonic lung fibroblasts. *Virology* 31:732–734.

16. Asano, Y. and M. Takahashi. 1978. Studies on neutralization of varicella-zoster virus and serological follow-up of cases of varicella and zoster. *Biken J.* 21:15–23.

17. Takahashi, M., T. Otsuka, Y. Okuno, Y. Asano, T. Yazaki, and S. Isomura. 1974. Live vaccine used to prevent the spread of varicella in children in hospital. *Lancet* 2:1288–1290.

18. Plotkin, S.A., J.D. Farquhar, M. Katz, and F. Buser. 1969. Attenuation of RA 27/3 rubella virus in WI-38 human diploid cells. *Am. J. Dis. Child.* 118:178–185.

19. Sabin, A.B. 1969. Biologic control of live attenuated rubella virus vaccines: Discussion. *Am. J. Dis. Child.* 118:378–379.

20. Hayflick, L., S. Plotkin, and R. E. Stevenson. 1987. History of the acceptance of human diploid cell strains as substrates for human virus vaccine manufacture. *Dev. Biol. Stand.* 68:9–17.

21. Hilleman, M.R., E.B. Buynak, J.E. Whitman, R.W. Weibel, and J. Stokes, Jr. 1969. Live attenuated rubella virus vaccines: Experiences with duck embryo cell preparations. *Am. J. Dis. Child.* 118:166–171.

22. Meyer, H.M., P.D. Parkman, T.E. Hobbins, and H.E. Larson. 1969. Attenuated rubella viruses: Laboratory and clinical characteristics. *Am. J. Dis. Child.* 118:155–165.

23. Prinzie, A., C. Huygelen, J. Gold, J.D. Farquhar, and J. Mckee. 1969. Experimental live attenuated rubella virus vaccine. *Am. J. Dis. Child.* 118:172–177.

24. Plotkin, S.A. 1994. Rubella vaccine. pp. 303–336. *In:* Plotkin S.A. and E.A. Mortimer, Jr. (Eds.), *Vaccines*, W.B. Saunders, Philadelphia.

25. Takahashi, M., Y. Asano, H. Kamiya, K. Baba, and K. Yamanishi. 1981. Active immunization for varicella-zoster virus. pp. 414–431. *In:* Nahmias A.J., W.R. Dowdle, and R.F. Schinazi (Eds.), *The Human Herpes Viruses: An Interdisciplinary Perspective*, Elsevier, New York.

26. Hayakawa, Y., S. Torigoe, K. Shiraki, K. Yamanishi, and M. Takahashi. 1984. Biological and biophysical markers of a live varicella vaccine strain (Oka): Identification of clinical isolates from vaccine recipients. *J. Infect. Dis.* 149:956–963.

27. Jacobs, J.P. 1976. The status of human diploid cell strain MRC-5 as an approved substrate for the production of viral vaccines. *J. Biol. Stand.* 4:97–99.

28. Asano, Y., H. Nakayama, T. Yazaki, R. Kato, S. Hirose, K. Tsuzuki, S. Ito, S. Isomura, and M. Takahashi. 1977. Protection against varicella in family contacts by immediate inoculation with live varicella vaccine. *Pediatrics* 59:3–7.

29. Asano, Y., H. Nakayama, T. Yazaki, S. Ito, S. Isomura, and M. Takahashi. 1977. Protective efficacy of vaccination in children in four episodes of natural varicella and zoster in the ward. *Pediatrics* 59:8–12.

30. Asano, Y. and M. Takahashi. 1977. Clinical and serologic testing of a live varicella vaccine and two-year follow-up for immunity of the vaccinated children. *Pediatrics* 60:810–814.

31. Baba, K., H. Yabuuchi, H. Okuni, and M. Takahashi. 1978. Studies with live varicella vaccine and inactivated skin test antigen: Protective effect of the vaccine and clinical application of the skin test. *Pediatrics* 61:550–555.

32. Hattori, A., T. Ihara, T. Iwasa, H. Kamiya, M. Sakurai, T. Izawa, and M. Takahashi. 1976. Use of live varicella vaccine in children with acute leukemia or other malignancies. *Lancet* 2:210.

33. Izawa, T., T. Ihara, A. Hattori, T. Iwasa, H. Kamiya, M. Sakurai, and M. Takahashi. 1977. Application of a live varicella vaccine in children with acute leukemia or other malignant diseases. *Pediatrics* 60:805–809.

34. Brunell, P.A. 1977. Commentary: Protection against varicella. *Pediatrics* 59:1–2.

35. Sabin, A.B. 1977. Commentary: Varicella-zoster virus vaccine. *JAMA* 238:1731–1733.

36. Plotkin, S.A. 1977. Varicella vendetta: Plotkin's plug. *Pediatrics* 59:953–954.

37. Brunell, P.A. 1977. Brunell's brush-off. *Pediatrics* 59:954.

38. Kempe, C.H. and A.A. Gershon. 1977. Commentary: Varicella vaccine at the crossroads. *Pediatrics* 60:930–931.

39. Gershon, A.A., S.P. Steinberg, L. Gelb, G. Galasso, W. Borkowski, P. LaRussa, A. Ferrera, and NIAID varicella vaccine collaborative study group. 1984. Live varicella vaccine: Efficacy for children with leukemia in remission. *JAMA* 252:355–362.

40. Lawrence, R., A.A. Gershon, R. Holzman, S.P. Steinberg, and NIAID varicella vaccine collaborative study group. 1988. The risk of zoster after varicella vaccination in children with leukemia. *N. Engl. J. Med.* 318:543–548.

41. Gershon, A.A., S. Steinberg, and NIAID collaborative varicella vaccine study group. 1989. Persistence of immunity to varicella in children with leukemia immunized with live attenuated varicella vaccine. *N. Engl. J. Med.* 320:892–897.

42. Tsolia, M., A. Gershon, S. Steinberg, and NIAID collaborative study group. 1990. Live attenuated varicella vaccine: Evidence that the virus is attenuated and the importance of skin lesions in transmission of varicella-zoster virus. *J. Pediatr.* 116:184–189.

43. Brunell, P.A., Z. Shehab, C. Geiser, and J.E. Waugh. 1982. Administration of live varicella vaccine to children with leukemia. *Lancet* 2:1069–1073.

44. Arbeter, A., S. Starr, R.E. Weibel, and S.A. Plotkin. 1982. Live attenuated varicella vaccine: Immunization of healthy children with the Oka strain. *J. Pediatr.* 100:886–893.

45. Weibel, R., B.J. Neff, B.J. Kuter, H.A. Guess, C.A. Rothenberger, A.J. Fitzgerald, K.A. Connor, A.A. McLean, M.R. Hilleman, and F.B. Buynak. 1984. Live at-

tenuated varicella virus vaccine: Efficacy trial in healthy children. *N. Engl. J. Med.* 310:1409–1415.

46. Arbeter, A.M., S.E. Starr, S. Preblud, T. Ihara, T. Paciorek, D.S. Miller, C.M. Zelson, E.A. Proctor, and S. Plotkin. 1984. Varicella vaccine follow-up studies. *Am. J. Dis. Child.* 138:434–438.

47. Diaz, P.S., D. Au, S. Smith, M. Amylon, M. Link, S. Smith, and A.M. Arvin. 1991. Lack of transmission of the live attenuated varicella vaccine to immunocompromised children after immunization of their siblings. *Pediatrics* 87:166–170.

48. World Health Organization. 1985. WHO Expert Committee on Biological Standardization. Requirements for Varicella Vaccine (Live) (Requirements for Biological Substances No. 36), Geneva, WHO, pp. 102–133.

49. Haas, R.J., B. Belohradsky, R. Dickerhoff, K. Eichinger, R. Eife, H. Holtman, O. Goetz, U. Graubner, and P. Peller. 1985. Active immunization against varicella of children with acute leukemic or other malignancies on maintenance chemotherapy. *Postgrad. Med. J.* 61:69–72.

50. Heller, L., G. Berglund, L. Ahstrom, K. Hellstrand, and B. Wahren. 1985. Early results of a trial of the Oka-strain varicella vaccine in children with leukemic or other malignancies in Sweden. *Postgrad. Med. J.* 61:79–83.

51. Slordahl, S.H., D. Wiger, T. Stromoy, M. Degre, E. Thorsby, and O. Lie. 1985. Vaccination of children with malignant disease against varicella. *Postgrad. Med. J.* 61:85–92.

52. Austgulen, R. 1985. Immunization of children with malignant diseases with the Oka-strain varicella vaccine. *Postgrad. Med. J.* 61:93–95

53. Ninane, J., D. Latinne, M.T. Heremans-Bracke, M. DeBruyere, and G. Cornu. 1985. Live varicella vaccine in severely immunodepressed children. *Postgrad. Med. J.* 61:97–102.

54. Broyer, M. and B. Boudailliez. 1985. Varicella vaccine in children with chronic renal insufficiency. *Postgrad. Med. J.* 61:103–106.

55. Heath, R.B. and J.S. Malpas. 1985. Experience with the live Oka-strain varicella vaccine in children with solid tumors. *Postgrad. Med. J.* 61:107–111.

56. Andre, F.E. 1985. Worldwide experience with the Oka-strain live varicella vaccine. *Postgrad. Med. J.* 61:113–120.

57. Just, M., R. Berger, and D. Luescher. 1985. Live varicella vaccine in healthy individuals. *Postgrad. Med. J.* 61:129–132.

58. Berger, R., D. Luescher, and M. Just. 1985. Restoration of varicella-zoster virus cell-mediated immune response after varicella booster immunization. *Postgrad. Med. J.* 61:143–145.

59. Duchateau, J., R. Vrijens, J. Nicaise, E. D'Hondt, H. Bogearts, and F.E. Andre. 1985. Stimulation of specific immune response to varicella antigens in the elderly with varicella vaccine. *Postgrad. Med. J.* 61:147–150.

60. Plotkin, S.A., A.A. Arbeter, and S.E. Starr. 1985. The future of varicella vaccine. *Postgrad. Med. J.* 61:155–162.

61. Ndumbe, P.M., J.E. Cradock-Watson, S. MacQueen, H. Bunn, F. Andre, E.G Davies, J.A. Dudgeon, and R.J. Levinsky. 1985. Immunization of nurses with a live varicella vaccine. *Lancet* 1:1144–1147.

62. Ozaki, T., T. Ichikawa, Y. Matsui, T. Nagai, Y. Asano, K. Yamanishi, and M. Takahashi. 1984. Viremic phase in nonimmunocompromised children with varicella. *J. Pediatr.* 104:85–87.

63. Asano, Y., N. Itakura, H. Yabuuchi, S. Hiroishi, T. Nagai, T. Ozaki, T. Yazaki, K. Yamanishi, and M. Takahashi. 1985. Viremia is present in incubation period in nonimmunocompromised children with varicella. *J. Pediatr.* 106:69–71.

64. Ozaki, T., T. Ichikawa, Y. Tatsui, K. Kondo, T. Nagai, Y. Asano, K. Yamanishi, and M. Takahashi. 1986. Lymphocyte-associated viremia in varicella. *J. Med. Virol.* 19:249–253.

65. Asano, Y., N. Itakura, Y. Hiroishi, S. Hirose, T. Ozaki, T. Kuno, T. Nagai, T. Yazaki, K. Yamanishi, and M. Takahashi. 1985. Viral replication and immunologic responses in children naturally infected with varicella-zoster virus and in varicella vaccine recipients. *J. Infect. Dis.* 152:863–868.

66. Matsunaga, Y., K. Yamanishi, and M. Takahashi. 1982. Experimental infection and immune response of guinea pigs with varicella-zoster virus. *Infect. Immun.* 37:407–412.

67. Kurata, T., R. Hondo, S. Sato, A. Oda, Y. Aoyama, and J.B. McCormick. 1983. Detection of viral antigens in formalin fixed specimens by enzyme treatment. *Ann. N.Y. Acad. Sci.* 420:192–207.

68. Takahashi, M. 1992. Current status and prospects of live varicella vaccine. *Vaccine* 10:1007–1014.

69. Hope-Simpson, R.E. 1965. The nature of herpes zoster: A long term study and a new hypothesis. *Proc. Roy. Soc. Med.* 58:9–20.

70. Kamiya, H., T. Kato, M. Isaji, K. Oitani, M. Ito, T. Ihara, M. Sakurai, and M. Takahashi. 1984. Immunization of acute leukemic children with a live varicella vaccine. *Biken J.* 27:99–102.

71. Yabuuchi, H., H. Baba, N. Tsuda, S. Okada, O. Nose, T. Seino, K. Tomita, K. Ha, T. Mimaki, M. Ogawa, T. Kanesaki, M. Yoshida, and M. Takahashi. 1984. A live varicella vaccine in a pediatric community. *Biken J.* 27:43–49.

72. Brunell, P.A., J. Taylor-Wiedeman, C.F. Geiser, L. Frieson, and E. Lydick. 1986. Risk of herpes zoster in children with leukemia: Varicella vaccine compared with history of chickenpox. *Pediatrics* 77:53–56.

73. Takahashi, M. 1986. Varicella vaccine. *Pediatric Clinics* (in Japanese) 49:1941–1949.

74. Takahashi, M. 1986. Development and characterization of varicella vaccine. *Clinical Virology* (in Japanese) 14:74–79.

75. Takahashi, M. 1987. A vaccine to prevent chickenpox. pp. 179–209. *In:* Hyman R.W. (Ed.), *Natural History of Varicella-Zoster Virus*, CRC Press, Boca Raton, FL.

76. Takahashi, M., K. Baba, K. Horiuchi, H. Kamiya, and Y Asano. 1990. A live varicella vaccine. *Adv. Exp. Med. Biol.* 278:49–58.

77. Hardy, I.B., A.A. Gershon, S.P. Steinberg, P. LaRussa, and Varicella Vaccine Collaborative Study Group. 1991. The incidence of zoster after immunization with live attenuated vaccine. A study in children with leukemia. *N. Engl. J. Med.* 325:1545–1550.

78. Johnson, R.T. and C.A. Mims. 1968. Pathogenesis of viral infections of the nervous system. *N. Engl. J. Med.* 278:23–30.

Acronyms and Abbreviations

AALAC	American Association for Laboratory Animal Care
ABSL	Animal Biosafety Levels
ALL	acute lymphocytic leukemia
ATTC	American Type Tissue Culture
AUTM	Association of University Technology Managers
BLA	biologics license application
CagA	cytotoxin-associated antigen
CBER	Center for Biologics Evaluation and Research
CD	cluster of differentiation
CDC	Center for Disease Control and Prevention
CDER	Center for Drugs Evaluation and Research
CFR	Code of Federal Regulations
CFU	colony-forming units
CIP	Continuation-In-Part
CMI	cell-mediated immunity
CRADA	Cooperative Research and Development Agreement
CTA	clinical trial agreement
DNA	deoxyribonucleic acid
DVRPA	Division of Vaccines and Related Products Applications
EGF	epidermal growth factor
EIA	enzyme-linked immunosorbent assay
ELA	establishment license application
FDA	Food and Drug Administration
FLC	Federal Laboratory Consortium
FTE	full-time equivalent
FTTA	Federal Technology Transfer Act
GATT	General Agreement on Tariffs and Trade
GLPs	good laboratory practices
GMP	good manufacturing practices
GOGO	government owned/government operated
GPEF	guinea pig embryo fibroblasts
HBIG	Hepatitis B immune globulin
HEL	human embryonic lung
HIV	human immunodeficiency virus
IB	investigator's brochure

ICH	International Conference on Harmonization
ID	infectious-dose
Igs	immunoglobulins
ILAR	Institute of Laboratory Animal Resources
IND	investigational new drug
IRB	institutional review board
LCM	lymphocyticchoriomeningitis
LEMSIP	Laboratory for Experimental Medicine and Surgery
MHC	major histocompatibility complex
MHV	mouse hepatitis virus
MSDS	material safety data sheet
MTAs	Material Transfer Agreement
MVM	minute virus of mice
NDA	new drug application
NIH	National Institutes of Health
NTIS	National Technical Information Service
OSHA	Occupational Safety and Health Administration
PBMCs	peripheral-blood mononuclear cells
PHS	Public Health Service
PI	principal investigator
PLA	product license application
PT	pertussis toxin
PTC	points to consider
PTO	[U.S.] Patent and Trademark Office
R&D	research and development
RPRCP	Regional Primate Research Centers Program
RSV	respiratory syncytial virus
SBIR	Small Business Innovation Research
SCID	severe combined immunodeficiency
SN	serial number
SOP	standard operating procedure
SPF	specific pathogen-free
STTR	Small Business Technology Transfer Research
TB	*M. tuberculosis*
TT	tetanus toxoid
UBMTA	Uniform Biological Material Transfer Agreement
USP	United States Pharmacopeia
VAF	virus antibody-free
VZV	varicella-zoster virus

Index

9 780367 400378